2010

Indiana Media
Directory

BRACKEMYRE PUBLISHING

Celebrating 23 Years of Publishing Media Directories

BRACKEMYRE PUBLISHING

10133 Preston Court, Fishers IN 46037

317/913-1655 (phone)

317/598-2609 (fax)

mediadirectory@comcast.net

www.MyMediaDirectory.com

Publisher: Lori Brackemyre
Assistant Publisher: Teresa Proffitt

Printed in the United States of America.

International Standard Book Number 978-0-944369-19-7

CONTENTS

NOTES

Brackemyre Publishing is celebrating its 23rd year publishing media directories, and we are pleased to present you with our latest edition—the 2010 Indiana Media Directory! The following notes may help you use the directory more efficiently:

1. Our directory is organized alphabetically by city and town. Under each city/town listing, news services, publications, radio stations and television stations are listed in that order. Publications which target a specific audience (college campus, business, etc.) are listed by category in our Specialty Publications section. All news services are also listed together in our News Services section.

2. When a publication or broadcast station has a mailing address in one town and a street address in another, the listing usually appears in the town of the mailing address.

3. A publication or broadcast station serves the city or town in which it is listed unless otherwise noted.

4. Fax numbers and e-mail addresses are listed if a media outlet has these and allows its number/address to be published.

5. If a media outlet has a telephone number specifically for news or a fax number specifically for news or advertising, those numbers are listed separately from the general numbers.

6. If a media outlet has an e-mail address specifically for news releases, then that e-mail address is listed separately from the general e-mail address.

7. Many media outlets use anti-spam filters to preclude them from receiving unsolicited e-mails. Hence, you may have e-mail messages returned to you as "undeliverable" that had correct e-mail addresses.

8. Of all the information in the directory, e-mail addresses are the most likely to change because of personnel changes or because a media outlet has changed its e-mail server. All e-mail addresses were verified and tested to be accurate when this directory was created.

9. All circulation numbers are approximate and rounded to the nearest hundred.

10. When targeting the media in one particular city, be sure to look up media in nearby cities or adjacent counties. There may be media which serve a city but are located outside of a city's limit. Use the County-City Index to find all of the cities and towns in a specific county that have media listings.

11. All publications accept advertising unless otherwise noted.

12. National publications which have offices in Indiana (but do not specifically serve Indiana) are not listed.

13. Television channel numbers listed are digital. Stations that are only available via local cable are noted as such.

14. All information has been obtained in writing or by telephone by Brackemyre Publishing.

ALBION

Albion New Era
Noble County

Mailing Address	P.O. Box 25, Albion IN 46701
Street Address	407 S. Orange St., Albion IN 46701
Telephone	260-636-2727
Telephone (toll-free)	877-636-2727
Fax	260-636-2042
E-mail	newera@app-printing.com
Web Site	www.app-printing.com
Publication Date	Weekly (Wednesday)
Circulation	2,000 (paid)
Publishing Company	All Printing & Publications
Publisher	Robert Allman	bob@app-printing.com
Editor	Joy LeCount	jlecount@app-printing.com
Advertising Manager	Barbara Crozier	ads@app-printing.com
Sports Editor	Jenny Gaff	cougarsports@app-printing.com

ALEXANDRIA

Alexandria Times-Tribune
Madison County

Mailing Address	P.O. Box 330, Alexandria IN 46001
Street Address	One Harrison Square, Alexandria IN 46001
Telephone	765-724-4469
Fax	765-724-4460
E-mail (news)	alextribune@elwoodpublishing.com
Web Site	www.elwoodpublishing.com
Publication Date	Weekly (Wednesday)
Circulation	2,000 (paid)
Publishing Company	Elwood Publishing Co.
Publisher	Robert Nash
Editor	Bill Fouts
Advertising Manager	Cindy Tyner

ANDERSON

The Herald-Bulletin — Madison Magazine — Pendleton News

Mailing Address	P.O. Box 1090, Anderson IN 46015
Street Address	1133 Jackson St., Anderson IN 46016
Telephone	765-622-1212
Telephone (toll-free)	800-750-5049
Telephone (news)	765-640-4800
Fax (news)	765-640-4815
Fax (advertising)	765-640-4820
E-mail (news)	newsroom@heraldbulletin.com
Web Site	www.theheraldbulletin.com
Publishing Company	Community Newspaper Holdings Inc.
Publisher	Henry Bird	hbird@cnhi.com
Managing Editor	Scott Underwood	scott.underwood@heraldbulletin.com
Assistant Managing Editor	Steve Dick	steve.dick@heraldbulletin.com
Advertising Manager	Connie Alexander	connie.alexander@heraldbulletin.com
Sports Editor	Rick Teverbaugh	rick.teverbaugh@heraldbulletin.com
Circulation Director	Amy Winter	amy.winter@heraldbulletin.com

The Herald-Bulletin

Publication Date	Daily (Sunday-Saturday)
Circulation	20,900-paid (daily); 22,100-paid (Sunday)

Madison Magazine

E-mail	madison@indianamediagroup.com
Publication Date	Quarterly
Circulation	6,000
Editor	Scott Miley	scott.miley@heraldbulletin.com

Pendleton News

E-mail	pendletonnews@heraldbulletin.com
Publication Date	Weekly (Thursday)
Circulation	4,100
Editor	Janis Bowling	janis.bowling@heraldbulletin.com
Notes	Serves Pendleton.

See **WBSB** (Muncie).

WFOF (90.3 FM) — **WGNR** (1470 AM) — **WGNR** (97.9 FM) — **WHPL** (89.9 FM) **WIWC** (91.7 FM) — **WMBL** (88.1 FM)

Address	1920 W. 53rd St., Anderson IN 46013
Telephone	765-642-2750
Telephone (toll-free)	888-877-9467
Fax	765-642-4033
E-mail	wgnr@moody.edu
E-mail (news)	wgnrnews@moody.edu
Web Site	www.wgnr.fm
Wattage	19,000 WFOF)
	1,000 (WGNR AM)
	50,000 (WGNR FM)
	2,000 (WHPL)
	2,100 (WIWC)
Format	Religious
On-air Hours	24/7
Broadcast Company	Moody Broadcasting
General Manager	Ray Hashley rhashley@moody.edu
News Director	Linda Yeager lyeager@moody.edu
Program Director	Tom Winn tom.winn@moody.edu
Accepts PSAs?	yes (contact wgnr@moody.edu)
Notes	Non-commercial stations. All stations except WGNR AM are a simulcast of WGNR FM. WFOF serves Covington. WGNR AM & FM serve Indianapolis. WHPL serves Lafayette. WIWC serves Kokomo. WMBL serves Mitchell.

See **WHBU, WMQX,** and **WMXQ** (Muncie).

See **WMDH FM** (New Castle).

WQME (98.7 FM)

Mailing Address	1100 E. 5th St., Anderson IN 46012
Street Address	1102 E. 6th St., Anderson IN 46012
Telephone	765-641-4349
Telephone (toll-free)	866-987-WQME
Telephone (news)	765-641-3804
Fax	765-641-3825
E-mail	email@wqme.com
E-mail (news)	news@wqme.com
Web Site	www.wqme.com
Wattage	6,000
Format	Contemporary Christian
Network Affiliations	CNN News
On-air Hours	24/7
Owner	Anderson University
General Manager	Donald Boggs gm@wqme.com
News/Program Director	Matt Rust mrust@wqme.com
Sales Manager	Gerald Longenbaugh gerald@wqme.com
Traffic Director	Norma Armogum business@wqme.com
Accepts PSAs?	yes (contact news@wqme.com)

See **WRFM FM** (Anderson).

See **WXFN** (Muncie).

ANGOLA

Herald-Republican
Steuben County

Address	45 S. Public Square, Angola IN 46703
Telephone	260-665-3117
Fax (news)	260-665-2322
E-mail	mikem@kpcnews.net
Web Site	www.heraldrepublicanonline.com
Publication Date	Daily (Sunday-Saturday)
Circulation	5,200 (paid)
Publishing Company	KPC Media Group
Publisher	Terry Housholder terryh@kpcnews.net
Editor	Mike Marturello mikem@kpcnews.net
Advertising Manager	Art Condon artc@kpcnews.net
Sports Editor	Ken Fillmore kenf@kpcnews.net
Notes	KPC Media Group main office is in Kendallville.

WEAX (88.3 FM)

Address	1 University Ave., Angola IN 46703
Telephone	260-665-4288
E-mail	hornbacherj@trine.edu
Web Site	www.88xradio.com
Wattage	920
Format	Alternative
On-air Hours	24/7
Owner	Trine University
General Manager	Josh Hornbacher hornbacherj@trine.edu
Sports Director	Aaron Coyle coylea@trine.edu
Notes	Non-commercial station. Does not broadcast local News.

WLKI (100.3 FM)

Mailing Address	P.O. Box 999, Angola IN 46703
Street Address	2655 N. State Road 127, Angola IN 46703
Telephone	260-665-9554
Fax	260-665-9064
E-mail	wlki@wlki.com
Web Site	www.wlki.com
Wattage	6,000
Format	Hot Adult Contemporary
Network Affiliations	Fox News
On-air Hours	24/7
Broadcast Company	Swick Broadcasting
General Manager	Steve Swick steve@wlki.com
News/Sports Director	Jim Measel news@wlki.com
Program Director	Andy St. John andy@wlki.com
Sales Manager	Bill Kerner bill@wlki.com
PSA Director	Wendy Kellett wendy@wlki.com
Accepts PSAs?	yes

ARCADIA

Hamilton County

See **WJCF** (Greenfield).

ATTICA

Fountain County Neighbor

Fountain County

Address	113 S. Perry St., Attica IN 47918
Telephone	765-762-2411
Fax	765-762-1547
E-mail	atticaeditor@sbcglobal.net
Web Site	www.fountaincountyneighbor.com
Publication Date	Weekly (Monday)
Circulation	2,100 (paid)
Publishing Company	Kankakee Valley Publishing Co.
Publisher	Don Hurd
Editor	Jennifer Baldwin atticaeditor@sbcglobal.net
Advertising Manager	Greg Willhite atticasales@sbcglobal.net
TMC/Shopper	Messenger (weekly)

WFWR (91.5 FM)

Address	909 S. McDonald St., Attica IN 47918
Telephone/Fax	765-764-1934
E-mail	wfwr91.5@comcast.net
Web Site	www.atticonline.com/wfwr.htm
Wattage	165
Format	Variety
Network Affiliations	SRN News
On-air Hours	24/7
Broadcast Company	Fountain Warren Community Radio
General Manager	John Dill
Engineer	Larry Grant
IT Information Officer	Dave Huckelberry
Accepts PSAs?	yes

AUBURN

DeKalb County

Butler Bulletin — Evening Star — Garrett Clipper

Address	118 W. 9th St., Auburn IN 46706
Telephone	260-925-2611
Fax	260-925-2625
Web Site	www.kpcnews.com
Publishing Company	KPC Media Group
Publisher	Terry Housholder terryh@kpcnews.net
Sports Editor	Mark Murdock markm@kpcnews.net
Business Editor	Kathryn Bassett kathrynb@kpcnews.net
Notes	KPC Media Group main office is in Kendallville.

Butler Bulletin

Mailing Address	P.O. Box 39, Butler IN 46721
Telephone	260-868-5501
E-mail (news)	jeffj@kpcnews.net
Publication Date	Weekly (Tuesday)
Circulation	1,500 (paid)
Editor	Jeff Jones jeffj@kpcnews.net
Advertising Manager	Art Condon artc@kpcnews.net
Notes	Serves Butler, St. Joe, and Spencerville.

Evening Star

E-mail (news)	dkurtz@kpcnews.net
Publication Date	Daily (Sunday-Saturday)
Circulation	9,500 (paid)
Editor	Dave Kurtz dkurtz@kpcnews.net
Advertising Manager	Art Condon artc@kpcnews.net

Garrett Clipper

Mailing Address	P.O. Box 59, Garrett IN 46738
Telephone	260-357-4123
E-mail (news)	suec@kpcnews.net
Publication Date	Semi-weekly (Monday & Thursday)
Circulation	1,200 (paid)
Editor	Sue Carpenter suec@kpcnews.net
Advertising Manager	Karen Bloom karenb@kpcnews.net
Notes	Serves Garrett.

WFGA (106.7 FM)

Address	450 N. Grandstaff Dr., Auburn IN 46706
Telephone	260-920-3602
Telephone (toll-free)	866-363-3764
Fax	260-920-3604
E-mail	woodrow@ilovefroggy.com
E-mail (news)	taylor@ilovefroggy.com
Web Site	www.ilovefroggy.com
Wattage	6,000
Format	Hot Adult Contemporary
Network Affiliations	Fox
On-air Hours	24/7
Broadcast Company	Fallen Timbers
General Mgr./Sales Mgr.	Woody Zimmerman woodrow@ilovefroggy.com
News Director	Taylor Brooks taylor@ilovefroggy.com
Promotions Director	Maggie Johnson maggie@ilovefroggy.com
Accepts PSAs?	yes (contact Jeff DeWeese)

WGLL (1570 AM)

Mailing Address	P.O. Box 11, Auburn IN 46706
Street Address	5446 County Rd. 29, Auburn IN 46706
Telephone	260-908-4984
Fax	260-387-6914
E-mail	tv7@comcast.net
Wattage	500
Format	Religious Talk
Network Affiliations	3ABN
On-air Hours	24/7
Owner	Raymond S. and Dorothy N. Moore Foundation
General Manager	Ray Alexander
Accepts PSAs?	yes (contact Ray Alexander)
Notes	Does not broadcast local news.

W07CL TV (Channel 26) — WFWC TV (Channel 10)

Mailing Address	P.O. Box 11, Auburn IN 46706
Street Address	5446 County Rd. 29, Auburn IN 46706
Telephone	260-908-4984
Fax	260-387-6914
E-mail	TV7@comcast.net
Network Affiliation	3ABN
On-air Hours	24/7
Broadcast Company	Raymond S. and Dorothy N. Moore Foundation
General Manager	Ray Alexander
Accepts PSAs?	yes (contact Ray Alexander)
Notes	Stations do not broadcast local news. WFWC TV serves Fort Wayne.

AURORA

Dearborn County

Journal-Press

Mailing Address	P.O. Box 59, Aurora IN 47001	
Street Address	414 Third St., Aurora IN 47001	
Telephone	812-926-0063	
Fax	812-926-0066	
E-mail	editor@registerpublications.com	
Web Site	www.thejournal-press.com	
Publication Date	Weekly (Tuesday)	
Circulation	4,700 (paid)	
Publishing Company	Register Publications	
Publisher	Joe Awad	editor@registerpublications.com
Editor	Erika Russell	community@registerpublications.com
Advertising Manager	Loretta Day	lday@registerpublications.com
Sports Editor	Jim Buchberger	sports@registerpublications.com
TMC/Shopper	The Marketplace (weekly)	
Notes	Main office at Register Publications (Lawrenceburg).	

AUSTIN

Scott County

See **Scott County Journal/Chronicle** (Pekin).

AVILLA

Noble County

Kendallville Mall

Mailing Address	P.O. Box 313, Avilla IN 46710
Street Address	109 Baum St., Avilla IN 46710
Telephone	260-897-2674
Fax	260-897-2697
Publishing Company	Scher Maihem Publishing
Publisher	Julia Scher
E-mail	mallnooz@embarqmail.com
Web Site	www.kendallvillemall.com
Publication Date	Monthly
Circulation	14,000 (free/mailed & newsstand)
Notes	Serves Kendallville, Avilla, Rome City & Wolcottville.

AVON

Hendricks County

Hendricks County Flyer — Westside Flyer

Address	8109 Kingston St., STE 500, Avon IN 46123	
Telephone	317-272-5800	
Telephone (toll-free)	800-359-3747	
Fax	317-272-5887	
E-mail (news)	kathy.linton@flyergroup.com	
Web Site	www.flyergroup.com	
Publication Date	Semi-weekly: Wednesday & Saturday (Hendricks County Flyer)	
	Weekly: Tuesday (Westside Flyer)	
Circulation	43,000-free/delivered (Hendricks County Flyer)	
	11,500-free/delivered (Westside Flyer)	
Publishing Company	cnhi media	
Publisher	Harold Allen	harold.allen@flyergroup.com
Editor	Kathy Linton	kathy.linton@flyergroup.com
Advertising Manager	Bill Jarchow	bill.jarchow@indianamediagroup.com
Sports Editor	Todd Taylor	todd.taylor@flyergroup.com
Notes	Westside Flyer serves west side of Indianapolis.	

See **Hendricks County ICON** (Plainfield).

Indianapolis Star—West Bureau

Address	8217 Kingston St., Avon IN 46123
Telephone	317-444-2800
Fax	317-444-8800
E-mail	westbureau@indystar.com
Web Site	www.indystar.com
Notes	Main office in Indianapolis.

BATESVILLE

Ripley County

The Herald-Tribune

Address	475 Huntersville Rd., Batesville IN 47006
Telephone	812-934-4343
Fax	812-934-6406
Web Site	www.batesvilleheraldtribune.com
Publication Date	Semi-weekly (Tuesday & Friday)
Circulation	3,200 (paid)
Publishing Company	Community Newspaper Holdings Inc.
Publisher	Laura Welborn	laura.welborn@indianamediagroup.com
Editor	Bryan Helvie	bryan.helvie@batesvilleheraldtribune.com
Advertising Manager	Keith Wells	keith.wells@indianamediagroup.com

WRBI (103.9 FM)

Mailing Address	P.O. Box 201, Batesville IN 47006
Street Address	133 S. Main St., Batesville IN 47006
Telephone	812-934-5111
Telephone (news)	812-934-5112
Fax	812-934-2765
E-mail	wrbi@wrbiradio.com
E-mail (news)	marymattingly@wrbiradio.com
Web Site	www.wrbiradio.com
Wattage	3,000
Format	Country
Network Affiliations	CNN
On-air Hours	24/7
Broadcast Company	White River Broadcasting Corp.
General Manager	Ronald E. Green	rongreen@wrbiradio.com
News/Public Affairs Dir.	Mary Mattingly	marymattingly@wrbiradio.com
Program/Sports Director	Caz Burdette	cazburdette@wrbiradio.com
Promotions Dir./Office Mgr.	Barbara Nolting	barbnolting@wrbiradio.com
Accepts PSAs?	yes (contact Caz Burdette)	

See **WYGS** (Columbus).

BEDFORD

Lawrence County

Times-Mail

Mailing Address	P.O. Box 849, Bedford IN 47421
Street Address	813 16th St., Bedford IN 47421
Telephone	812-275-3355
Telephone (toll-free)	800-782-4405
Fax (news)	812-277-3472
Fax (advertising)	812-275-4191
Web Site	www.tmnews.com
Publication Date	Daily (Sunday-Saturday)
Circulation	13,500 (paid)
Publishing Company	Schurz Communications
Publisher/Editor	Mayer Maloney mmaloney@heraldt.com
Managing Editor	Mike Lewis mikel@tmnews.com
Advertising Manager	Cory Bollinger cbollinger@heraldt.com
Sports Editor	Sean Duncan sduncan@tmnews.com
TMC/Shopper	Hoosier Shopper (weekly)
Notes	**Hoosier Times** (Sunday edition) printed in conjunction with The Herald Times (Bloomington).

WBIW (1340 AM) — WPHZ (102.5 FM) — WQRK (105.5 FM)

Mailing Address	P.O. Box 1307, Bedford IN 47421
Street Address	424 Heltonville Rd., Bedford IN 47421
Telephone	812-275-7555
Fax	812-279-8046
E-mail (news)	news@wbiw.com
Network Affiliations	Fox News
On-air Hours	24/7
General Mgr./Promo Dir.	Holly Davis hlindsey@hpcisp.com
News/Public Affairs Dir.	Gage Lutes news@wbiw.com
Program/Sports Director	Jason Thompson jason@wbiw.com
Sales Manager	Sarah Reinhard sarah@superoldies.net
Traffic Manager	Tammy Farley tammy@superoldies.net
Accepts PSAs?	yes (contact Holly Davis)

WBIW

Web Site	www.wbiw.com
Wattage	1,000
Format	News/Talk
Broadcast Company	Ad-Venture Media, Inc.

WPHZ

E-mail	comments@wphz.com
Web Site	www.wphz.com
Wattage	6,000
Format	Adult Contemporary
Broadcast Company	Mitchell Community Broadcast Co.
Notes	Serves Mitchell.

WQRK

E-mail	comments@superoldies.net
Web Site	www.superoldies.net
Wattage	2,000
Format	Oldies
Broadcast Company	Ad-Venture Media, Inc.

BEECH GROVE

Southside Times
Marion County

Address	301 Main St., Beech Grove IN 46107
Telephone	317-787-3291
Fax	317-787-3325
E-mail (news)	news@ss-times.com
Web Site	www.ss-times.com
Publication Date	Weekly (Thursday)
Circulation	21,500 (free/delivered & newsstand)
Publishing Company	Times-Leader Publications
Publisher	Roger Huntzinger	rhuntzin@ss-times.com
Editor	Sara Gentry	news@ss-times.com

BERNE

Adams County

Berne Tri-Weekly News

Address	153 S. Jefferson St., Berne IN 46711
Telephone	260-589-2101
Fax	260-589-8614
E-mail (news)	news@bernetriweekly.com
Web Site	www.bernetriweekly.com
Publication Date	Semi-weekly (Monday, Wednesday & Friday)
Circulation	2,000 (paid)
Publishing Company	DRG Indiana
Publisher	Roger Muselman
Managing Editor	Clint Anderson

WZBD (92.7 FM)

Mailing Address	P.O. Box 4050, Berne IN 46711
Street Address	955 Hwy. 27 N., Berne IN 46711
Telephone	260-589-9300
Fax	260-589-8045
Fax (sales)	260-724-9002
E-mail	wzbd@onlyinternet.net
Web Site	www.wzbd.com
Wattage	6,000
Format	Full Service Adult Contemporary
On-air Hours	24/7
Broadcast Company	Adams County Radio Inc.
General Manager	Robert Weaver
News Director	Bob Adams
Operations Director	Tony Giltner
Sales Manager	Al Converset
Accepts PSAs?	yes (contact Tony Giltner)

BICKNELL
See **WUZR** (Vincennes).

Knox County

BLOOMFIELD

Greene County

See **Greene County Daily World** (Linton).

BLOOMINGTON

Monroe County

Bloom Magazine

Mailing Address	P.O. Box 1204, Bloomington IN 47402
Street Address	209 N. Washington St., Bloomington IN 47408
Telephone	812-323-8959
Fax	812-323-8965
Web Site	www.magbloom.com
Publication Date	Bi-monthly
Circulation	13,000 (free & paid subscription)
Publishing Company	Bloomington Magazine, Inc.
Publisher/Editor	Malcolm Abrams editor@magbloom.com
Assoc. Publisher/Ad Mgr.	Erica De Santis erica@magbloom.com
Associate Editor	Ron Eid ron@magbloom.com
Notes	Culture/lifestyle publication. Also publishes **Our Town** (annual guide to Bloomington, www.OurTownBloomington.com)

Herald-Times

Mailing Address	P.O. Box 909, Bloomington IN 47402
Street Address	1900 S. Walnut St., Bloomington IN 47401
Telephone	812-332-4401
Telephone (toll-free)	800-422-0070
Fax (news)	812-331-4383
Fax (advertising)	812-331-4285
Web Site	www.HeraldTimesOnline.com
Publication Date	Daily (Sunday-Saturday)
Circulation	26,500-paid (daily); 43,200-paid (Sunday)
Publishing Company	Schurz Communications
Publisher	E. Mayer Maloney Jr. mmaloney@heraldt.com
Editor	Bob Zaltsberg rzaltsberg@heraldt.com
Managing Editor	Andrea Murray amurray@heraldt.com
Advertising Manager	Cory Bollinger cbollinger@heraldt.com
Sports Editor	Chris Korman ckorman@heraldt.com
Notes	**Hoosier Times** (Sunday newspaper) has three editions serving Bedford, Bloomington and Martinsville.

See **Indiana Daily Student** (listed under College Campus/Specialty Publications).

See **Inside Indiana** (listed under Sports/Specialty Publications).

Ryder Magazine

Address	504 W. 4th St., Bloomington IN 47404
Telephone/Fax	812-339-2002
E-mail	peter@theryder.com
Web Site	www.theryder.com
Publication Date	Monthly
Circulation	20,000 (free/newsstand)
Publishing Company	In the Dark Enterprises
Publisher/Editor	Peter LoPilato peter@theryder.com
Notes	Arts & entertainment publication

WBWB (96.7 FM) — WHCC (105.1 FM)

Address	304 S. State Road 446, Bloomington IN 47401
Telephone	812-336-8000
Fax	812-336-7000
On-air Hours	24/7
Broadcast Company	Artistic Media Partners
General Manager	Sandy Zehr sandy@artisticradio.com
Sales Manager	Edward Ice ed@wbwb.com
Accepts PSAs?	yes (submit via fax)

WBWB

E-mail	wbwb@wbwb.com
Web Site	www.wbwb.com
Wattage	3,000
Format	Top 40 Contemporary Hit Radio
Program Director	Brandon Scott brandon@wbwb.com
Promotions Director	Edward Ice ed@wbwb.com

WHCC

E-mail	whcc105@whcc105.com
Web Site	www.whcc105.com
Wattage	6,000
Format	Country
Networks	IU Sports Network, NASCAR
Program Director	Rick Evans rick@whcc105.com
Promotions Director	Sheila Stephen sheila@whcc105.com

See **AIR 1** (listed under National Radio Stations).

See **WCJL** (Valparaiso).

WCLS (97.7 FM)

Address	5858 W. Hwy. 46, Bloomington IN 47404
Telephone	812-935-7400
Telephone (toll-free)	877-337-0977
Fax	812-935-7404
E-mail	wclsfm@smithville.net
Web Site	www.wclsfm.com
Wattage	6,000
Format	Classic Hits
Network Affiliations	ABC
On-air Hours	24/7
Broadcast Company	Mid-America Radio
General Mgr./Sales Mgr.	David Bruce davidbruce@wclsclassichits.com
Program/Sports Director	Tony Kale wclsfm@smithville.net
Accepts PSAs?	yes (contact psa@wclsclassichits.com)

WFHB (91.3 FM)

Mailing Address	P.O. Box 1973, Bloomington IN 47402
Street Address	108 W. 4th St., Bloomington IN 47404
Telephone	812-323-1200
Fax	812-323-0320
E-mail (news)	news@wfhb.org
Web Site	www.wfhb.org
Wattage	1,600
Format	News/Public Affairs/Music
Network Affiliations	PRI
On-air Hours	24/7
Broadcast Company	Bloomington Community Radio, Inc.
General Manager	Will Murphy manager@wfhb.org
News/Public Affairs Dir.	Chad Carrothers news@wfhb.org
Music Director	Jim Manion ionman@wfhb.org
Accepts PSAs?	yes—text-only PSAs/no prerecorded PSAs (contact psa@wfhb.org)
Notes	Non-commercial station. Repeats on 98.1 FM serving south central Indiana; 100.7 FM serving Brown County; and 106.3 FM serving Ellettsville. All repeater frequencies are 250 watts.

WFIU (103.7 FM)

Address	1229 E. 7th St., Bloomington IN 47405
Telephone	812-855-1357
Fax	812-855-5600
E-mail	wfiu@indiana.edu
Web Site	www.wfiu.org
Wattage	35,000
Format	Classical/Jazz
Network Affiliations	NPR, PRI
On-air Hours	24/7
Owner	Indiana University Trustees
General Manager	Perry Metz
News Director	Stan Jastrzebski
Station Manager	Christina Kuzmych
Accepts PSAs?	yes
Notes	Non-commercial station. Repeats on100.7 FM (Columbus), 106.1 FM (Kokomo), 95.1 FM (Terre Haute), 101.7 FM (French Lick), and 98.9 FM (Greensburg).

WGCL (1370 AM & 95.9 FM) — WTTS (92.3 FM)

Address	400 One City Centre, Bloomington IN 47404
Telephone	812-332-3366
Telephone (news)	812-339-6397
Fax	812-331-4570
On-air Hours	24/7
Broadcast Company	Sarkes Tarzian Inc.
General Manager	Geoff Vargo	geoff@wttsfm.com
News/Program Director	Brad Holtz	brad@wttsfm.com
Accepts PSAs?	yes (contact Brad Holtz)

WGCL

E-mail (news)	news@wgclradio.com
Web Site	www.wgclradio.com
Wattage	5,000
Format	News/Talk
Networks	ABC
Sports Director	Joe Smith	smitty1370@yahoo.com
Sales Manager	Duncan Myers	duncan@wgclradio.com

WTTS

E-mail	brad@wttsfm.com
Web Site	www.wttsfm.com
Wattage	37,000
Format	Adult Alternative
Sales Manager	Daryl McIntire	mac@wttsfm.com
Promotions Director	Johnette Harvey	johnette@wttsfm.com
Notes	Sales office is in Indianapolis.

WIUX (99.1 FM)

Address	815 E. 8th St., Bloomington IN 47408
Telephone	812-855-7862
Fax	812-855-1073
Web Site	www.wiux.org
Wattage	22
Format	Indie
On-air Hours	24/7
Broadcast Company	Indiana University Student Broadcasting
Accepts PSAs?	yes
Notes	Non-commercial station. All positions filled by students.

WMYJ (88.9 FM) — WVNI (95.1 FM)

Mailing Address	P.O. Box 1628, Bloomington IN 47402
Street Address	4317 E. 3rd St., Bloomington IN 47401
Telephone	812-335-9500
Fax	812-335-8880
Network Affiliations	SRN, SMN
On-air Hours	24/7
Broadcast Company	Mid-America Corp.
General Manager	David Keister
Operations Manager	Jim Webster jim@spirit95fm.com
Program Director	Kyle Watson kyle@spirit95fm.com
Accepts PSAs?	yes (contact Kyle Watson)

WMYJ

E-mail	wmyj@spirit95fm.com
Web Site	www.spirit95fm.com/myjoy
Wattage	6,000
Format	Southern Gospel
Notes	Serves Bedford & Bloomington.

WVNI

E-mail	spirit95@spirit95fm.com
Web Site	www.spirit95fm.com
Wattage	4,200
Format	Contemporary Christian

Community Access Television Services

(Cable-only channels 3, 7, 12, 14 & 96)

Address	303 E. Kirkwood Ave., Bloomington IN 47408
Telephone	812-349-3111
Fax	812-349-3112
Web Site	www.catstv.net
Network Affiliation	independent
On-air Hours	24/7
General Manager	Michael White mbwhite@monroe.lib.in.us
Accepts PSAs?	yes (contact Adam Stillwell, stillz@hotmail.com)
Notes	Non-commercial station. Does not broadcast local news.
	Channel formats: 3 (library), 7 (public access), 12 (city government), 14 (county government), and 96 (scola).

See **WCLJ-TV** (Greenwood).

WTIU TV (Channels 30.1/TIU HD; 30.2/TIU World; 30.3/TIU Family; & 30.4/TIU Espanol)

Address	1229 E. 7th St., Bloomington IN 47405	
Telephone	812-855-5900	
Telephone (toll-free)	800-662-3311	
Telephone (news)	812-855-6200	
Fax	812-855-0729	
Fax (news)	812-855-1177	
E-mail	wtiu@indiana.edu	
E-mail (news)	aestrahl@indiana.edu	
Web Site	www.wtiu.indiana.edu	
Network Affiliation	PBS	
On-air Hours	24/7	
Owner	Indiana University	
General Manager	Perry Metz	metz@indiana.edu
Station Manager	Phil Meyer	pwmeyer@indiana.edu
News Director	Ann Shea	aestrahl@indiana.edu
General Sales Manager	Marianne Woodruff	mawoodru@indiana.edu
Program Director	Brent Molnar	brmolnar@indiana.edu
Promotions Director	Ann Wesley	amwesley@indiana.edu
Accepts PSAs?	no	
Notes	Non-commercial station.	

BLUFFTON

News-Banner

Wells County

Mailing Address	P.O. Box 436, Bluffton IN 46714	
Street Address	125 N. Johnson St., Bluffton IN 46714	
Telephone	260-824-0224	
Fax	260-824-0700	
E-mail	email@news-banner.com	
Web Site	www.news-banner.com	
Publication Date	Daily (Monday-Saturday)	
Circulation	5,100 (paid)	
Publishing Company	News-Banner Publications	
Publisher/Editor	Mark Miller	miller@news-banner.com
Managing Editor	Glen Werling	glenw@news-banner.com
Advertising Manager	Jean Bordner	jeanb@news-banner.com
Sports Editor	Paul Beitler	sports@news-banner.com
Assitant Editor	David Schultz	daves@news-banner.com
TMC/Shopper	The Echo/Sunriser News (weekly)	

BOONVILLE

Boonville Standard — Newburgh Register

Warrick County

Mailing Address	P.O. Box 266, Boonville IN 47601
Street Address	204 W. Locust St., Boonville IN 47601
Telephone	812-897-2330
Fax	812-897-3703
E-mail	newsroom@warricknews.com
Web Site	www.warricknews.com
Publishing Company	Warrick Publishing
Publisher	Gary Neal — gwneal@aol.com
Managing Editor	Wendy Wary — wwary@warricknews.com
TMC/Shopper	Warrick County Today (weekly)

Boonville Standard

Publication Date	Weekly (Thursday)
Circulation	4,000 (paid)

Newburgh Register

Publication Date	Weekly (Thursday)
Circulation	10,000 (paid)
Editor	Tim Young — tyoung@warricknews.com
Notes	Serves Newburgh.

WBNL (1540 AM)

Mailing Address	P.O. Box 270, Boonvile IN 47601
Street Address	2177 N. Hwy. 61, Boonville IN 47601
Telephone	812-897-2080
Telephone (news)	812-897-2081
Fax	812-897-2130
E-mail	rturpen@radio1540.net
Web Site	www.radio1540.net
Wattage	250
Format	Easy Listening
Network Affiliations	IRN, Learfield, Network Indiana
On-air Hours	24/7
Broadcast Company	Turpen Communications LLC
General Mgr./Sales Mgr.	Ralph Turpen — rturpen@radio1540.net
News/Sports Director	Larry Schweizer
Accepts PSAs?	yes (contact Carolyn Bryant, missbrooks@radio1540.net)

BOURBON

See **Bourbon News-Mirror** (Plymouth).

Marshall County

BRAZIL

Clay County

Brazil Times

Mailing Address	P.O. Box 429, Brazil IN 47834
Street Address	100 N. Meridian St., Brazil IN 47834
Telephone	812-446-2216
Telephone (toll-free)	800-489-5090
Fax	812-446-0938
E-mail (news)	news@thebraziltimes.com
Web Site	www.thebraziltimes.com
Publication Date	Daily (Monday and Wednesday-Saturday)
Circulation	4,700 (paid)
Publishing Company	Rust Communications
Publisher	Randy List	rlist2@hotmail.com
General Manager	Lynne Llewellyn	lynnellewellyn@yahoo.com
Managing Editor	Jason Moon	scoop1j@gmail.com
Sports Editor	Carey Fox	redwood17257@yahoo.com

See **WSDM** and **WSDX** (Terre Haute).

BREMEN

Marshall County

Advance News — Bremen Enquirer

Address	126 E. Plymouth St., Bremen IN 46506
Telephone	574-546-2941
Fax	574-546-5170
E-mail	enquirer@fourway.net
Web Site	www.thepilotnews.com
Publication Date	Weekly (Thursday)
Circulation	2,500-paid (Advance News)
	1,800-paid (Bremen Enquirer)
Publishing Company	Horizon Publications
Publisher	Rick Kreps	rkreps@thepilotnews.com
General Manager	Jerry Bingle	jbingle@thepilotnews.com
Editor	Mandy McFarland	enquirer@fourway.net
Managing Editor	Maggie Nixon	mnixon@thepilotnews.com
Advertising Manager	Cindy Stockton	cstockton@thepilotnews.com
TMC/Shopper	The Shopper (weekly)
Notes	Main office at The Pilot-News (Plymouth).
	Advance News serves Nappanee.

BRIGHT

Dearborn County

Bright Beacon

Address	23995 Stateline Rd., STE E, Bright IN 47025
Telephone	812-637-0660
Fax	812-637-5300
E-mail	brightbeacon1@fuse.net
Publication Date	Monthly
Circulation	5,300 (free/mailed)
Publishing Company	Beacon Publishing Inc.
Publisher	Celeste Calvitto

BRISTOL

Elkhart County

Bristol Bugle News

Mailing Address	P.O. Box 414, Bristol IN 46507
Street Address	16900 County Rd. 104, Bristol IN 46507
Telephone	574-848-1404
Fax	574-848-5689
E-mail	bristolnews@aol.com
Web Site	www.bristolbugle.com
Publication Date	Monthly
Circulation	4,600 (free/mailed & newsstand)
Publishing Company	Bristol News and Printing
Publisher/Editor	Laurie Eads
Managing Editor	James Eads
Advertising Manager	Laura Shaffer

BROOK

Newton County

See **Brook Reporter** (Kentland).

BROOKVILLE

Franklin County

Brookville American — Brookville Democrat

Mailing Address	P.O. Box 38, Brookville IN 47012	
Street Address	531 Main St., Brookville IN 47012	
Telephone	765-647-4221	
Fax	765-647-4811	
E-mail	john@thebrookvillenews.com	
Web Site	www.thebrookvillenews.com	
Publication Date	Weekly (Wednesday)	
Circulation	1,300-paid (Brookville American)	
	4,700-paid (Brookville Democrat)	
Publishing Company	Whitewater Publications	
Publisher	Gary Wolf	gary@thebrookvillenews.com
Editor	John Estridge	john@thebrookvillenews.com
Advertising Manager	Dawn Grizzell	ads@thebrookvillenews.com
Sports Editor	Andy Sallee	andy@thebrookvillenews.com

BROWNSBURG

Hendricks County

Brownsburg Week

Address	533 E. Main St., Brownsburg IN 46112
Telephone/Fax	317-852-7166
E-mail	brownsburgweek@sbcglobal.net
Web Site	www.brownsburgweek.com
Publication Date	Weekly (Monday)
Circulation	2,000 (free/newsstand)
Publisher/Editor	Bob Wilson

See **Hendricks County ICON** (Plainfield).

XRB (1610 AM)

Address	701 N. Green St., Brownsburg IN 46112	
Telephone	317-852-1610	
Fax	866-265-4900	
E-mail	shane.ray@radiobrownsburg.com	
Web Site	www.radiobrownsburg.com	
Wattage	1/10 watt	
Format	Oldies	
On-air Hours	24/7	
Owner/General Manager	Shane Ray	shane.ray@radiobrownsburg.com
Sales Manager	Sherry Moodie	sherry@radiobrownsburg.com
Accepts PSAs?	yes (contact Shane Ray)	

BROWNSTOWN

Jackson County

Jackson County Banner

Mailing Address	P.O. Box G, Brownstown IN 47220	
Street Address	116 E. Cross St., Brownston IN 47220	
Telephone	812-358-2111	
Fax	812-358-5606	
E-mail (news)	news@thebanner.com	
E-mail (advertising)	ads@thebanner.com	
E-mail (sports)	sports@thebanner.com	
Web Site	www.thebanner.com	
Publication Date	Semi-weekly (Tuesday & Thursday)	
Circulation	3,000 (paid)	
Publishing Company	Community Media Group	
Publisher	Patricia Robertson	probertson@thebanner.com
TMC/Shopper	The Weekly Budget (weekly)	

BUTLER

DeKalb County

See **Butler Bulletin** (Auburn).

CAMBRIDGE CITY

Wayne County

Nettle Creek Gazette — Western Wayne News

Mailing Address	P.O. Box 337, Cambridge City IN 47327
Street Address	26 W. Church St., Cambridge City IN 47327
Telephone	765-478-5448
Fax	765-478-5155
E-mail	nettlecreekgazette@verizon.net
	westernwaynenews@verizon.net
Publication Date	Weekly (Wednesday)
Circulation	1,100-paid (Nettle Creek Gazette)
	2,800-paid (Western Wayne News)
Owner/Publisher/Editor	Janis Buhl janisbuhl@verizon.net
Notes	Nettle Creek Gazette serves Hagerstown.
	Western Wayne News serves Cambridge City and Centerville.

CARMEL

Hamilton County

See **atCarmel Community Newsletter** (Indianapolis).

See **Carmel Business Leader** (listed under Business/Specialty Publications).

Current in Carmel — Current in Noblesville
Current in Westfield

Address	One S. Range Line Rd., STE 220, Carmel IN 46032
Telephone	317-489-4444
Fax	317-489-4446
E-mail (news)	info@currentincarmel.com
Web Site	www.youarecurrent.com
Publication Date	Weekly (Tuesday)
Circulation	28,500-free/mailed (Current in Carmel)
	24,000-free/mailed (Current in Noblesville)
	8,800-free/mailed (Current in Westfield)
Publishing Company	Current Publishing LLC
Publisher	Brian Kelly brian@currentincarmel.com
Executive Editor/Ad Mgr.	Steve Greenberg steve@currentincarmel.com
Managing Editor	Bryan Unruh bryan@currentincarmel.com
Notes	Current in Noblesville serves Noblesville. Current in Westfield serves Westfield.

WHJE (91.3 FM)

Address	520 E. Main St., Carmel IN 46032
Telephone	317-846-7721, ext. 7531
Fax	317-571-4066
E-mail	whje@whje.com
Web Site	www.whje.com
Wattage	400
Format	Triple A
Network Affiliations	Network Indiana
On-air Hours	24/7
Owner	Carmel-Clay Schools
General Manager	Brian Spilbeler bspilbel@ccs.k12.in.us
Accepts PSAs?	yes (contact Brian Spilbeler)
Notes	Non-commercial station.

WRDZ (98.3 FM)

Address	630 W. Carmel Dr., STE 160, Carmel IN 46032
Telephone	317-574-2000
Telephone (toll-free)	877-574-2001
Fax	317-581-1985
Web Site	www.radiodisney.com
Wattage	3,000
Format	Family Hits
Network Affiliations	Radio Disney
On-air Hours	24/7
Broadcast Company	Walt Disney Co.
General Manager	Jim McConville jim.mcconville@disney.com
Program Director	Ray De La Garza ray.delagarza@disney.com
Promotions Director	Laura Sanchez laura.a.sanchez@disney.com
Accepts PSAs?	yes (contact Jim McConville)
Notes	Does not broadcast local news.

CAYUGA
Vermillion County

The Herald News

Mailing Address	P.O. Box 158, Cayuga IN 47928
Street Address	103 E. Curtis St., Cayuga IN 47928
Telephone/Fax	765-492-4401
E-mail	heraldnews@sbcglobal.net
Web Site	www.challengernewspapers.com
Publication Date	Weekly (Wednesday)
Circulation	1,000 (paid/mailed)
Publishing Company	Greenwood Newspapers Inc.
Publisher	Doug Chambers doug@indychallenger.com
Editor	Diane Siddens heraldnews@sbcglobal.net

CEDAR LAKE
Lake County

See **Cedar Lake Journal** (Lowell).

See **Cedar Lake/Lowell Star** (Crown Point).

CENTERVILLE
Wayne County

See **Western Wayne News** (Cambridge City).

CHARLESTOWN
Clark County

Leader—Bureau

Address	382 Main Cross St., Charlestown IN 47111
Telephone/Fax	812-256-3377
Notes	Main office in Pekin.

CHESTERTON

Porter County

Chesterton Tribune

Mailing Address	P.O. Box 919, Chesterton IN 46304
Street Address	193 S. Calumet Rd., Chesterton IN 46304
Telephone	219-926-1131
Fax	219-926-6389
E-mail	news@chestertontribune.com
Web Site	www.chestertontribune.com
Publication Date	Daily (Monday-Friday)
Circulation	5,000 (paid)
Publisher	Warren Canright
Editor	David Canright
Sports Editor	T. R. Harlan

See **The Chronicle** (Valparaiso).

WBEW (89.5 FM)

Address	848 E. Grand Ave., Chicago IL 60611
Telephone	312-893-2956
Fax	312-948-4878
E-mail	info@vocalo.org
Web Site	www.vocalo.org
Wattage	4,000
Format	Talk/Eclectic
On-air Hours	24/7
General Manager	Wendy Turner wendy@vocalo.org
Accepts PSAs?	yes
Notes	Licensed to Chesterton. Serves Chicago, IL & Northwest Indiana. Sister station to WBEZ.

WBEZ (91.5 FM)—Northwest Indiana News Bureau

Address	442 N. Calumet Rd., Chesterton IN 46304
Telephone	219-926-6236
Fax	219-926-6386
Web Site	www.chicagopublicradio.org
Wattage	8,300
Format	Public Radio
On-air Hours	24/7
Chesterton Reporter	Michael Puente mpuente@chicagopublicradio.org
Accepts PSAs?	yes
Notes	Repeater station of Chicago Public Radio. Sister station to WBEW.

WDSO (88.3 FM)

Address	2125 S. 11th St., Chesterton IN 46304
Telephone	219-983-3777
Fax	219-983-3775
Web Site	www.wdso.org
Wattage	400
Format	Rock
Network Affiliations	Network Indiana
On-air Hours	24/7 (Monday 6:00 a.m. - Friday 6:00 p.m.)
Owner	Duneland School Corp.
General Manager	Matthew Waters matthew.waters@duneland.k12.in.us
Operations Manager	Michele Stipanovich michele.stipanovich@duneland.k12.in.us
Accepts PSAs?	yes
Notes	Non-commercial, educational station.

CHURUBUSCO

Churubusco News
Whitley County

Mailing Address	P.O. Box 8, Churubusco IN 46723
Street Address	123 N. Main St., Churubusco IN 46723
Telephone	260-693-3949
Fax	260-693-6545
E-mail	cheditor@app-printing.com
Web Site	www.app-printing.com
Publication Date	Weekly (Wednesday)
Circulation	2,200 (paid)
Publishing Company	All Printing & Publications
Publisher	Robert Allman
Editor/Advertising Mgr.	David Crabill

CICERO
Hamilton County

See **WJCY** (Valparaiso).

CLARKSVILLE
Clark County

Louisville Courier-Journal—Bureau

Address	2500 Lincoln Dr., Clarksville IN 47129
Telephone	812-948-1315
Fax	812-949-4041
E-mail	jtaylor@courier-journal.com
Web Site	www.courier-journal.com
Indiana Editor	Joe Taylor jtaylor@courier-journal.com
Notes	Main office in Louisville, KY (800-765-4011)

WNDA (1570 AM)

Mailing Address	P.O. Box 2623, Clarksville IN 47131
Street Address	220 Potters Lane, Clarksville IN 47129
Telephone	812-949-1570
Telephone (news)	812-949-0009
Fax	812-949-9632
Fax (news)	812-949-5056
E-mail	rocky@indiana1570.com
Web Site	www.kool1570.com
Wattage	1,500
Format	Oldies
Network Affiliations	Network Indiana
On-air Hours	24/7
Broadcast Company	New Albany Broadcasting
General Manager	David Smith david@indiana9.com
Program Director	Rocky Knight rocky@indiana1570.com
Sales Manager	Corissa Smith csmith@indiana9.com
Accepts PSAs?	yes (contact Rocky Knight)

WJYL TV
(Channels 16.1/TBN; 16.2/The Church Channel; 16.3 /JCTV & 16.4/Smile of a Child)

Mailing Address	P.O. Box 2605, Clarksville IN 47131
Street Address	515 Potters Lane, Clarksville IN 47129
Telephone	812-949-9595
Fax	812-949-5221
E-mail	prayer@wjyl.org
Web Site	www.wjyl.org
Network Affiliation	TBN
Broadcast Company	Dominion Media
General Manager	John Smith
Notes	Does not broadcast local news. Serves Clarksville, IN & Louisville, KY.

WNDA TV (Channel 9.1) — WYCS TV (Channel 24.1)

Mailing Address	P.O. Box 2623, Clarksville IN 47131
Street Address	220 Potters Lane, Clarksville IN 47129
Telephone	812-949-9843
Telephone (news)	812-949-0009
Fax	812-949-5056
E-mail (news)	news@indiana9.com
E-mail (sports)	sports@indiana9.com
Web Sites	www.indiana9.com
	www.wycstv.com
On-air Hours	24/7
Broadcast Company	New Albany Broadcasting
General Mgr./News Dir.	David Smith david@indiana9.com
General Sales Manager	Fred North fred@indiana9.com
Accepts PSAs?	yes

CLAY CITY

Clay County

Clay City News
Mailing Address	P.O. Box 38, Clay City IN 47841
Street Address	717 Main St., Clay City IN 47841
Telephone	812-939-2163
Fax	812-939-2286
E-mail	ccnews@claycitynews.com
Publication Date	Weekly (Wednesday)
Circulation	2,000 (paid)
Publishing Company	Spencer Evening World
General Manager	John A. Gillaspie

CLINTON

Vermillion County

Daily Clintonian
Mailing Address	P.O. Box 309, Clinton IN 47842
Street Address	422 S. Main St., Clinton IN 47842
Telephone	765-832-2443
E-mail	cccc@mikes.net
Web Site	www.ccc-clintonian.com
Publication Date	Daily (Monday-Friday)
Circulation	5,400 (paid)
Publishing Company	Clinton Color Crafters
Publisher	George B. Carey
Editor	Jinanne Carey
Advertising Manager	B. Falls
Sports Editor	Cathy Craig

See **WPFR FM** (Terre Haute).

CLOVERDALE

Putnam County

Hoosier Topics
Mailing Address	P.O. Box 496, Cloverdale IN 46120	
Street Address	1 N. Main St., Cloverdale IN 46120	
Telephone	765-795-4438	
Telephone (toll-free)	877-795-4438	
Fax	765-795-3121	
E-mail	htopics@ccrtc.com	
Web Site	www.countryconnect.com/htopics/htopics.html	
Publication Date	Weekly (Tuesday)	
Circulation	20,000 (free/mailed)	
Publisher	John Gillaspy	
Advertising Manager	Jenny Snyder	jsnyder@ccrtc.com

COLUMBIA CITY

Whitley County

Post & Mail

Mailing Address	P.O. Box 837, Columbia City IN 46725
Street Address	927 W. Connexion Way, Columbia City IN 46725
Telephone	260-244-5153
Fax	260-244-7598
E-mail	postandmail@earthlink.net
Web Site	www.thepostandmail.com
Publication Date	Daily (Monday-Saturday)
Circulation	4,000-paid (Monday & Wedneday-Saturday)
	13,600-free (Tuesday)
Publishing Company	Horizon Publications
Publisher	Doug Brown
Editor	Ruth Stanley
Advertising Manager	Mick Long
Sports Editor	Andrew Shultz
TMC/Shopper	weekly

Publisher — Doug Brown — dougbrown@thepostandmail.com
Editor — Ruth Stanley — ruth@thepostandmail.com
Advertising Manager — Mick Long — pmadvertising@earthlink.net
Sports Editor — Andrew Shultz — andrew@thepostandmail.com

WJHS (91.5 FM)

Address	600 N. Whitley St., Columbia City IN 46725
Telephone	260-248-8915
Fax	260-244-7326
Web Site	www.wjhs915.org
Wattage	2,650
Format	Adult Alternative
On-air Hours	24/7
Owner	Whitley County Consolidated Schools
General Manager	Krystal Walker-Zoltek
Program Director	Laurie Walls
Accepts PSAs?	yes (contact Laurie Walls)
Notes	Non-commercial station.

General Manager — Krystal Walker-Zoltek — walkerzoltekkd@wccs.k12.in.us
Program Director — Laurie Walls — wallsll@wccs.k12.in.us

COLUMBUS

See **Columbus Parent** (listed under Parenting/Specialty Publications).

Bartholomew County

The Republic
Address	333 Second St., Columbus IN 47201
Telephone	812-372-7811
Telephone (toll-free)	800-876-7811
Telephone (news)	812-379-5633
Fax	812-379-5706
Fax (news)	812-379-5711
Fax (advertising)	812-379-5776
E-mail (news)	editorial@therepublic.com
Web Site	www.therepublic.com
Publication Date	Daily (Sunday-Saturday)
Circulation	23,000-paid (daily); 24,000-paid (Sunday)
Publishing Company	Home News Enterprises
Publisher	Chuck Wells cwells@therepublic.com
Editor	Bob Gustin rlgustin@therepublic.com
Sports Editor	Tyler Hoeppner thoeppner@therepublic.com

See **K-Love** (listed under National Radio Stations).

WAUZ (89.1 FM) — WKRY (88.1 FM) — WYGS (91.1 FM)
Mailing Address	P.O. Box 2626, Columbus IN 47202
Street Address	825 Washington St., Columbus IN 47201
Telephone	812-375-9947
Telephone (toll-free)	800-603-9873
E-mail	ygs@wygs.org
Web Site	www.wygs.org
Wattage	800 (WAUZ)
	500 (WKRY)
	380 (WYGS)
Format	Christian/Gospel
On-air Hours	24/7
Broadcast Company	Good Shepherd Radio, Inc.
Executive Director	David Burnett dburnett@wygs.org
Accepts PSAs?	yes (contact David Burnett)
Notes	WAUZ (Greensburg) and WKRY (Versailles) are a simulcast of WYGS. WYGS repeats on 94.5 FM (Salem), 101.9 FM (Palmyra), 97.5 FM (Batesville), and 94.3 FM (LaGrange, KY). Non-commercial stations. Stations do not broadcast local news.

WCSI (1010 AM) — WINN (104.9 FM) — WKKG (101.5 FM) — WWWY (106.1 FM)

Mailing Address	P.O. Box 1789, Columbus IN 47202
Street Address	3212 Washington St., Columbus IN 47203
Telephone	812-372-4448
Telephone (news)	812-376-4770
Fax	812-372-1061
E-mail (news)	news@wcsiradio.com
Network Affiliations	Fox, Network Indiana
On-air Hours	24/7
Broadcast Company	White River Broadcasting Company
General Mgr./Sales Mgr.	Tasha Mann tasha@wcsiradio.com
News Director	Lani Weigler laniweigler@wcsiradio.com
Sports Director	Sam Simmermaker samsimmermaker@wcsiradio.com
Accepts PSAs?	yes (contact psa@wcsiradio.com)

WCSI

E-mail	wcsi@wcsiradio.com
Web Site	www.wcsiradio.com
Wattage	500
Format	News/Talk
Program Director	John Foster jfoster@wcsiradio.com

WINN

E-mail	studio@1049theriver.fm
Web Site	www.1049theriver.fm
Wattage	6,000
Format	Classic Hits
Program Director	Rich Anthony richanthony@1049theriver.fm

WKKG

E-mail	studio@wkkg.com
Web Site	www.wkkg.com
Wattage	50,000
Format	Country
Program Director	Scott Michaels scottmichaels@wkkg.com

WWWY

E-mail	rockme@y106.com
Web Site	www.y106.com
Wattage	50,000
Format	Rock
Program Director	Tonya Haze tonyahaze@y106.com

See **WFIU** (Bloomington).

WHUM (98.5 FM)

Address	1325 Washington St., STE D, Columbus IN 47201
Telephone	812-379-9985
E-mail	whumradio@gmail.com
Web Site	www.whum.org
Wattage	100
Format	Eclectic
On-air Hours	24/7
Broadcast Company	Columbus Community Radio Corp.
General Manager	Mitzi Quinn
Accepts PSAs?	yes (contact Mitzi Quinn)
Notes	Repeats on 98.3 FM (Seymour).

WRZQ (107.3 FM) — WYGB (100.3 FM)

Mailing Address	P.O. Box 690, Columbus IN 47202	
Street Address	825 Washington St., Columbus IN 47201	
Telephone	812-379-1077	
Fax	812-375-2555	
Network Affiliations	Hoosier Ag Today	
On-air Hours	24/7	
Broadcast Company	Reising Radio Partners	
General Manager	Mike King	mking@qmix.com
News Director	Brittany Gray	news@qmix.com
Sports Director	Tom Rust	saceit@face-2-face.org
Sales Manager	Dawn Andrews	ddaugherty@qmix.com
Promotions Director	Sara Beth Clark	sclark@qmix.com
Operations Manager	Dave Wineland	dwineland@qmix.com
Accepts PSAs?	yes (contact Brittany Gray)	

WRZQ

E-mail	qmix@qmix.com
E-mail (news)	news@qmix.com
Web Site	www.qmix.com
Wattage	25,000
Format	Adult Contemporary

WYGB

E-mail	korn@korncountry.com
E-mail (news)	news@qmix.com
Web Site	www.korncountry.com
Wattage	3,000
Format	Country

CONNERSVILLE

News-Examiner

Fayette County

Mailing Address	P.O. Box 287, Connersville IN 47331
Street Address	406 N. Central Ave., Connersville IN 47331
Telephone	765-825-0581
Telephone (toll-free)	888-906-1700
Fax	765-825-4599
E-mail	newsexaminer@newsexaminer.com
Web Site	www.newsexaminer.com
Publication Date	Daily (Monday-Saturday)
Circulation	7,900 (paid)
Publishing Company	Paxton Media Group
Publisher	Rachel Raney	rraney@shelbynews.com
General Manager/Ad Mgr.	Kelly Pierce	kpierce@newsexaminer.com
Editor	Gary Hufferd	ghufferd@newsexaminer.com
Sports Editor	Mike Moffett	mmoffett@newsexaminer.com
TMC/Shopper	Shopper Stopper (weekly)

WIFE (1580 AM) — WIFE-FM (94.3 FM)

Mailing Address	P.O. Box 619, Connersville IN 47331
Street Address	406-1/2 Central Ave., Connersville IN 47331
Telephone	765-825-6411
Telephone (toll-free)	866-843-9433
Fax	765-825-2411
E-mail (news)	news@wifefm.com
Network Affiliations	USA
On-air Hours	24/7
Broadcast Company	Rodgers Broadcasting
General Manager	Jerri Pruet	jerri@wifefm.com
News Director	Mike Selke	selke@wifefm.com
Program Director	Ted Cramer	ted@wifefm.com
Sales Manager	Steve Frey	steve@g1013.com
Accepts PSAs?	yes (contact Jerri Pruet)

WIFE

Web Site	www.superoldies1580.com
Wattage	250
Format	Super Oldies

WIFE-FM

Web Site	www.wifefm.com
Wattage	1,500
Format	Country
Notes	Additional office in Rushville.

See **WJCF** and **WRFM FM** (Greenfield).

TV3 (Cable-only Channel 3)

Address	500 N. Central Ave., Connersville IN 47331
Telephone	765-825-3245
Fax	765-827-3135
E-mail	citytv@connersvillein.gov
Web Site	www.connersvillecommunity.com
On-air Hours	24/7
Owner	City of Connersville
General Manager	John Pause	citytv@connersvillein.gov
Sports Director	Ben Houston
Program Director	Justin Roberts	jroberts@connersvillein.gov
Accepts PSAs?	yes (contact John Pause)
Notes	Non-commercial station. Government/education access channel.

CORYDON

Harrison County

Clarion News — Corydon Democrat

Address	301 N. Capitol Ave., Corydon IN 47112
Fax	812-738-1909
Publication Date	Weekly (Wednesday)
Publishing Company	O'Bannon Publishing Co.
Publisher	Jonathan O'Bannon	jobannon@corydondemocrat.com
Advertising Manager	Karen Hanger	ads@corydondemocrat.com
TMC/Shopper	The Shopper (weekly)

Clarion News

Telephone	812-738-4552
E-mail	cadams@clarionnews.net
Web Site	www.clarionnews.net
Circulation	17,600 (free/mailed & delivered)
Editor	Chris Adams	cadams@clarionnews.net
Notes	Serves English.

Corydon Democrat

Telephone	812-738-2211
E-mail	jsaylor@corydondemocrat.com
Web Site	www.corydondemocrat.com
Circulation	8,700 (paid)
Editor	Jo Ann Spieth-Saylor	jsaylor@corydondemocrat.com
Sports Editor	Brian Smith	bsmith@corydondemocrat.com

WOCC (1550 AM)

Mailing Address	P.O. Box 838, Corydon IN 47112
Street Address	211 N. Capitol Ave., Corydon IN 47112
Telephone	812-738-9622
Fax	812-738-1676
E-mail	wocc1550@cs.com
Web Site	www.woccam1550.com
Wattage	250
Format	Classic Oldies
Network Affiliations	Network Indiana, USA
On-air Hours	24/7
Owner/General Manager	Richard Brabandt
Sports Director	Chris Stoner
Public Affairs/Promotions	Mary Ann Brabandt
Accepts PSAs?	yes (contact Mary Brabandt)

COVINGTON

Fountain County

See **WFOF** (Anderson).

WKZS (103.1 FM) — WSKL (92.9 FM)

Mailing Address	P.O. Box 67, Danville IL 61834	
Street Address	820 Railroad St., Covington IN 47932	
Fax	765-793-4644	
Network Affiliations	Jones	
On-air Hours	24/7	
General Manager	Rhea Weatherford	rhea@kisscountryradio.com
News Director	Gregory Green	greg@kisscountryradio.com
Program Director	Larry Weatherford	lar@kisscountryradio.com
Promotions Director	Tara Auter	tara@kisscountryradio.com
Accepts PSAs?	yes (contact Tara Auter)	

WKZS

Telephone	765-793-5477
E-mail	info@kisscountryradio.com
Web Site	www.kisscountryradio.com
Wattage	3,000
Format	Country
Broadcast Company	Benton-Weatherford Broadcasting

WSKL

Telephone	765-793-4823
E-mail	fmkool929@aol.com
Web Site	www.koololdies.net
Wattage	4,500
Format	Oldies
Broadcast Company	Zona Communications

CRAWFORDSVILLE

Montgomery County

See **Bachelor** (listed under College Campus/Specialty Publications).

Journal Review

Mailing Address	P.O. Box 512, Crawfordsville IN 47933
Street Address	119 N. Green St., Crawfordsville IN 47933
Telephone	765-362-1200
Telephone (toll-free)	800-488-4414
Telephone (news)	765-362-1201
Fax	765-364-5427
Fax (news)	765-364-5424
Fax (advertising)	765-364-5425
E-mail (news)	jheater@jrpress.com
Web Site	www.journalreview.com
Publication Date	Daily (Monday-Saturday)
Circulation	8,600 (paid)
Publishing Company	PTS, Inc.
Publisher	Sean Smith	ssmith@jrpress.com
General Manager	Shawn Storie	sstorie@jrpress.com
Executive Editor	Jay Heater	jheater@jrpress.com
Sports Editor	Matt Wilson	mwilson@jrpress.com
TMC/Shopper	Advertiser

Paper of Montgomery County

Mailing Address	P.O. Box 272, Crawfordsville IN 47933
Street Address	101 W. Main St., STE 300, Crawfordsville IN 47933
Telephone	765-361-0100
Fax (news)	765-361-2945
Fax (advertising)	765-361-1882
E-mail (news)	news@thepaper24-7.com
Web Site	www.thepaper24-7.com
Publication Date	Daily (Monday-Saturday)
Circulation	6,800 (paid)
Publisher	Tim Timmons
Managing Editor	Barry Lewis
TMC/Shopper	Weekly of West Central Indiana (weekly)

WCDQ (106.3 FM) — **WCVL** (1550 AM) — **WIMC** (103.9 FM)

Mailing Address	P.O. Box 603, Crawfordsville IN 47933
Street Address	1757 N. 175 W., Crawfordsville IN 47933
Telephone	765-362-8200
Fax	765-364-1550
E-mail	dapeach@forchtbroadcasting.com
Web Site	www.crawfordsvilleradio.com
Wattage	3,400 (WCDQ)
	250 (WCVL)
	1,350 (WIMC)
Format	Country (WCDQ)
	Adult Standards/Nostalgia (WCVL)
	Classic Hits (WIMC)
Network Affiliations	Jones, Brownfield, MRN (WCDQ)
	ABC News, Jones, Brownfield (WCVL)
	ABC News (WIMC)
On-air Hours	24/7
Broadcast Company	Forcht Broadcasting
General Manager	Dave Peach dapeach@forchtbroadcasting.com
News Director	Mark Webber mawebber@forchtbroadcasting.com
Accepts PSAs?	yes (contact Mark Webber)
Notes	Broadcasts news from the Paper of Montgomery County.

WNDY (91.3 FM)

Address	301 W. Wabash Ave., Crawfordsville IN 47933
Telephone	765-361-6240
Fax	765-361-6424
Wattage	2,200
Format	College Format/NPR
On-air Hours	24/7
Owner	Wabash College
Faculty Advisor	Brent Harris harrisb@wabash.edu
Notes	Simulcast of WFYI FM, Indianapolis (3:00 a.m. to 6:00 p.m.).
	College format (6:00 p.m. to 3:00 a.m.).

WVRG (93.9 FM)

Address	915 N. Whitlock Ave., Crawfordsville IN 47933
Telephone	765-364-0158
E-mail	wvrglp@sbcglobal.net
Wattage	87
Format	Bible Teaching/Contemporary Christian Music
On-air Hours	24/7
Owner	Calvary Chapel of Crawfordsville
Accepts PSAs?	yes
Notes	Non-commercial station.

CROTHERSVILLE

Crothersville Times

Mailing Address	P.O. Box 141, Crothersville IN 47229
Street Address	510 Moore St., STE 100, Crothersville IN 47229
Telephone/Fax	812-793-2188
E-mail	ctimes@crothersville.net
Web Site	www.crothersvilletimes.com
Publication Date	Weekly (Wednesday)
Circulation	1,400 (paid)
Publisher/Editor	Curt Kovener

See **WOJC** (Valparaiso).

CROWN POINT

Cedar Lake/Lowell Star — Crown Point Star

Mailing Address	P.O. Box 419, Crown Point IN 46308	
Street Address	112 W. Clark St., Crown Point IN 46307	
Telephone	219-663-4212	
Fax	219-663-0137	
Publication Date	Weekly (Thursday)	
Circulation	2,200 (paid)	
Publishing Company	Sun Times Media Group	
Managing Editor	Kim Mathisen	kmathisen@post-trib.com
Sports Editor	Mark Smith	msmith@post-trib.com
TMC/Shopper	Shopping News (weekly)	
Notes	Cedar Lake/Lowell Star serves Cedar Lake and Lowell.	

The Times of Northwest Indiana—Bureau

Address	2080 N. Main St., Crown Point IN 46307	
Telephone	219-662-5300	
Fax	219-662-5295	
Lake County Editor	Sharon Ross	sharon.ross@nwitimes.com
Notes	Main office in Munster.	

CULVER

See **Culver Citizen** (Plymouth).

DALE

Spencer County Leader
Spencer County

Mailing Address	P.O. Box 206, Dale IN 47523
Street Address	218 E. Medcalf St., Dale IN 47523
Telephone	812-937-2100
Fax	812-367-2371
E-mail	thenews@psci.net
Publication Date	Weekly (Thursday)
Circulation	2,000 (paid)
Publishing Company	Dubois-Spencer Counties Publishing
Publishers	Richard & Kathy Tretter; Miriam & Paul Ash
Editor	Kathy Tretter	ferdnews@psci.net
Managing Editor	Cheryl Hurst	sclreporter@psci.net
Sports Editor	Brian Bohne	dssports@psci.net
Notes	Main office at Ferdinand News (Ferdinand).

DANVILLE

Republican
Hendricks County

Mailing Address	P.O. Box 149, Danville IN 46122
Street Address	6 E. Main St., Danville IN 46122
Telephone/Fax	317-745-2777
E-mail	therepublican@sbcglobal.net
Publication Date	Weekly (Thursday)
Circulation	1,600 (paid)
Publishing Company	Hendricks County Republican, Inc.
Publisher/Editor	Betty Jean Weesner

DECATUR

Decatur Daily Democrat
Adams County

Address	141 S. 2nd St., Decatur IN 46733
Telephone	260-724-2121
Fax	260-724-7981
E-mail (news)	editorial@decaturdailydemocrat.com
E-mail (advertising)	advertising@decaturdailydemocrat.com
Web Site	www.decaturdailydemocrat.com
Publication Date	Daily (Monday-Saturday)
Circulation	5,800 (paid)
Publishing Company	Horizon Publications
Publisher/Advertising Mgr.	Ron Storey	publisher@decaturdailydemocrat.com
Editor	Bob Shraluka	bob@decaturdailydemocrat.com
Sports Editor	Joe Spaulding	sports@decaturdailydemocrat.com
TMC/Shopper	Berne Shopping News

WADM (1540 AM)

Mailing Address	P.O. Box 530, Decatur IN 46733
Telephone	260-724-7161
Fax	260-724-8719
E-mail	jay@wadm.com
Web Site	www.wadm.com
Wattage	250
Format	Country Classics
Network Affiliations	Network Indiana, USA Radio
On-air Hours	daytime
Broadcast Company	Lewis Broadcasting LLC
General Manager	Jay Lewis jaylewis@lewisbroadcasting.com
Sales Manager	Kathy Lewis kathy@wadm.com
Accepts PSAs?	yes (contact Jay Lewis)
Notes	Does not broadcast local news.

DELPHI

Carroll County

Carroll County Comet—Delphi Office

Address	114 E. Franklin St., Delphi IN 46923
Telephone	765-564-2222
Fax	765-564-2010
E-mail	delphinews@carrollcountycomet.com
Web Site	www.carrollcountycomet.com
Notes	Main office in Flora.

DeMOTTE

Jasper County

Kankakee Valley Post-News

Mailing Address	P.O. Box 110, DeMotte IN 46310
Street Address	827 S. Halleck St., DeMotte IN 46310
Telephone	219-987-5111
Telephone (toll-free)	888-809-5561
Fax	219-987-5119
E-mail	kvpost@netnitco.net
Web Site	www.kvonline.info
Publication Date	Weekly (Thursday)
Circulation	2,800 (paid)
Publishing Company	Community Media Group
Publisher	Don Hurd
Executive Editor	Clayton Doty
Editor	Cindy Ward
Advertising Manager	Anita Padgett
TMC/Shopper	Action Plus Shopper (weekly)

DUNKIRK

Jay County

News and Sun

Mailing Address	P.O. Box 59, Dunkirk IN 47336
Street Address	209 S. Main St., Dunkirk IN 47336
Telephone	765-768-6022
Fax	260-726-8143
E-mail	cr.news@comcast.net
Web Site	www.thecr.com
Publication Date	Weekly (Wednesday)
Circulation	1,500 (paid)
Publishing Company	Graphic Printing Co.
Publisher	Jack Ronald	jack.ronald@comcast.net
Editor	Robert Banser Jr.	cr.news@comcast.net
Advertising Manager	Marlene Giddings	cr.ads@comcast.net
TMC/Shopper	The Circulator (weekly)
Notes	Main office at The Commericial Review (Portland).

DYER

Lake County

Shopper

Address	924 E. 162nd St., South Holland IL 60473
Telephone	219-836-1960
Telephone (toll-free)	800-410-5250
Fax	708-333-9630
E-mail	general@myshopper.biz
Web Site	www.myshopper.biz
Publication Date	Weekly (Wednesday)
Circulation	17,500 (free/mailed & deliverd)
Publisher/Editor	Arlo Kallemeyn
Notes	Serves Dyer, Schererville and St. John.
	Also publishes 4 other editions, which serve areas of Illinois.

EDINBURGH

Johnson County

See **Atterbury Crier** and **Edinburgh Courier** (Franklin).

See **WRFM FM** (Greenfield).

ELKHART

The Truth

Mailing Address	P.O. Box 487, Elkhart IN 46515
Street Address	421 S. 2nd St., Elkhart IN 46516
Telephone	574-294-1661
Telephone (toll-free)	800-585-5416
Telephone (news)	574-389-3630
Fax	574-294-4014
Fax (news)	574-294-3895
Fax (advertising)	574-293-3302
E-mail (news)	newsroom@etruth.com
Web Site	www.etruth.com
Publication Date	Daily (Sunday-Saturday)
Circulation	29,000 (paid)
Publishing Company	Truth Publishing Co
Owner	John Dille
Publisher/General Mgr.	Brandon Erlacher	berlacher@etruth.com
Managing Editor	Gregory Halling	ghalling@etruth.com
Sports Editor	Bill Beck	bbeck@etruth.com
TMC/Shopper	Smart Shopper (weekly)

WCMR (1270 AM) — WFRI (100.1 FM) — WFRN (104.7 FM) — WFRR (93.7 FM)

Mailing Address	P.O. Box 307, Elkhart IN 46515
Street Address	25802 County Rd. 26, Elkhart IN 46517
Telephone	574-875-5166
Telephone (toll-free)	800-933-0501
Telephone (news)	574-875-1500
Fax	574-875-6662
E-mail	wfrn@wfrn.com
E-mail (news)	news@wfrn.com
E-mail (sales)	sales@wfrn.com
Network Affiliations	IRN News, Network Indiana
On-air Hours	24/7
Broadcast Company	Progressive Broadcasting System Inc. (WCMR, WFRI, & WFRN)
	Christian Friends Broadcasting Inc. (WFRR)
General Mgr./Sales Mgr.	Ed Moore	emoore@wfrn.com
News/Sports Director	Don Wagner	news@wfrn.com
Promotions Director	Rachelle Brandt	rbrandt@wfrn.com

WCMR

Web Site	www.solidgospel1270.com
Wattage	5,000
Format	Southern Gospel/Conservative Talk
Program Director	Doug Moore	dmoore@wfrn.com
Accepts PSAs?	no

WFRI — WFRN — WFRR

Web Site	www.WFRN.com
Wattage	6,000 (WFRI & WFRR)
	50,000 (WFRN)
Format	Christian Contemporary
Program Director	Ed Moore	emoore@wfrn.com
Accepts PSAs?	yes (contact Gregg Richards, events @wfrn.com)
Notes	WFRN repeats on 95.1 FM (Michigan City), 96.1 FM (South Bend), 96.5 FM (Mishawaka), 106.7 FM (Valparaiso), and 107.7 FM (Marion). WFRI (Winamac) and WFRR (Kokomo) are a partial simulcast of WFRN.

WLEG (102.7 FM) — WTRC (1340 AM)

Address	421 S. 2nd St., Elkhart IN 46516
Telephone	574-389-5100
Telephone (news)	574-293-6397
Fax	574-389-5101
E-mail	kmusselman@federatedmedia.com
E-mail (news)	news@hippieradio1340.com
On-air Hours	24/7
Broadcast Company	Federated Media
General Manager	Brad Williams bwilliams@federatedmedia.com
News Director	Paul Weaver pweaver@hippieradio1340.com
Station/Sales Manager	Kevin Musselman kmusselman@federatedmedia.com
Accepts PSAs?	yes (contact Paul Weaver)

WLEG

Web Site	www.ilovemyfroggy.com
Wattage	3,000
Format	Hot Adult Contemporary
Network	ABC
Program & Public Affairs Director	Beau Kennedy bkennedy@federatedmedia.com

WTRC

Web Site	www.hippieradio1340.com
Wattage	1,000
Format	Music
Network	Fox
Program & Public Affairs Director	Paul Weaver pweaver@hippieradio1340.com

WLFQ (98.7 FM)

Address	2601 Benham Ave., Elkhart IN 46517
Telephone	574-295-4357
Fax	574-295-4673
E-mail	jt@lff.net
Web Site	www.wlfq.org
Wattage	100
Format	Christian/Rap/Hip Hop
On-air Hours	24/7
Owner	Living Faith Fellowship
General Manager	James Taylor jt@lff.net
Accepts PSAs?	yes
Notes	Does not broadcast local news.

WVPE (88.1 FM)

Address	2424 California Rd., Elkhart IN 46514
Telephone	574-262-5660
Telephone (toll-free)	888-399-9873
Fax	574-262-5700
E-mail	wvpe@wvpe.org
Web Site	www.wvpe.org
Wattage	11,500
Format	News/Talk/Jazz
On-air Hours	24/7
Owner	Elkhart Community School Corporation
General Manager	Anthony Hunt	ahunt@wvpe.org
Program Director	Lee Burdorf	lburdorf@wvpe.org
Development Director	Kim Macon	kmacon@wvpe.org
Accepts PSAs?	yes (contact Lee Burdorf)
Notes	Non-commercial station.

See **WZOW** (South Bend).

WNIT TV (Channels 34.1/HD & 34.2/Standard Definition)

Mailing Address	P.O. Box 3434, Elkhart IN 46515
Street Address	317 W. Franklin St, Elkhart IN 46516
Telephone	574-675-9648
Fax	574-262-8497
E-mail	wnit@wnit.org
Web Site	www.wnit.org
Network Affiliation	PBS
On-air Hours	24/7
Broadcast Company	Michiana Public Broadcasting Corp.
General Manager	Mary Pruess	mpruess@wnit.org
Local Sales Manager	Jody Freid	jfreid@wnit.org
Program Director	Angela Moisenko	amoisenko@wnit.org
Public Affairs Director	Doug Farmwald	dfarmwald@wnit.org
Promotions Director	Roger Chrastil	rchrastil@wnit.org
Accepts PSAs?	no
Notes	Non-commercial station. Does not broadcast local news.

ELLETTSVILLE

Monroe County

Ellettsville Journal

Mailing Address	P.O. Box 98, Ellettsville IN 47429
Street Address	211 N. Sale St., Ellettsville IN 47429
Telephone	812-876-2254
Fax	812-876-2853
E-mail	journal@bluemarble.net
Publication Date	Weekly (Wednesday)
Circulation	2,000 (paid)
Publisher	John Gillaspy
Editor	Travis Curry
Managing Editor	Steve Sturgeon scribe@bluemarble.net
Advertising Manager	Ranee Love
TMC/Shopper	Ellettsville Extra (weekly)

See **WFHB** (Bloomington).

ELWOOD

Madison County

Call-Leader

Mailing Address	P.O. Box 85, Elwood IN 46036
Street Address	317 S. Anderson St., Elwood IN 46036
Telephone	765-552-3355
Fax	765-552-3358
E-mail	elpub@elwoodpublishing.com
E-mail (news)	elwoodeditor@elwoodpublishing.com
Web Site	www.elwoodpublishing.com
Publication Date	Daily (Monday-Saturday)
Circulation	3,100 (paid)
Publishing Company	Elwood Publishing Co.
Publisher	Robert Nash
Editor	Sandy Burton
Advertising Manager	Jay Putterbaugh
Sports Editor	Ed Hamilton
TMC/Shopper	Leader-Tribune Review East (weekly)

ENGLISH

Crawford County

See **Clarion News** (Corydon).

EVANSVILLE

Vanderburgh County

See **Crescent Magazine** (listed under College Campus/Specialty Publications).

See **Evansville Business** (listed under Business/Specialty Publications).

Evansville Courier & Press

Mailing Address	P.O. Box 268, Evansville IN 47702	
Street Address	300 E. Walnut St., Evansville IN 47713	
Telephone	812-424-7711	
Telephone (toll-free)	800-288-3200	
Telephone (news)	812-464-7412	
Fax (news)	812-422-8196	
Fax (advertising)	812-464-7487	
E-mail	courier@courierpress.com	
Web Site	www.courierpress.com	
Publication Date	Daily (Sunday-Saturday)	
Circulation	70,000-paid (daily); 90,000-paid (Sunday)	
Publishing Company	Evansville Courier Co.	
Publisher	Jack Pate	patej@courierpress.com
Editor	Mizell Stewart III	stewartm@courierpress.com
Managing Editor	Linda Negro	negrol@courierpress.com
Asst. Managing Ed./News	Kathleen Wagner	wagnerk@courierpress.com
Asst. Managing Ed./Visuals	Kevin Swank	swankk@courierpress.com
Advertising Manager	David Hedge	hedged@courierpress.com
Sports Editor	Tim Ethridge	ethridget@courierpress.com
Business Editor	Tom Lovett	lovett@courierpress.com
Online Editor	Ryan Reynolds	reynoldsr@courierpress.com
TMC/Shopper	Extra (weekly)	
Notes	Bureau in Indianapolis.	

Evansville Living

Address	223 N.W. 2nd St., STE 200, Evansville IN 47708	
Telephone	812-426-2115	
Fax	812-426-2134	
E-mail	jennifer@evansvilleliving.com	
Web Site	www.evansvilleliving.com	
Publication Date	Bi-monthly	
Circulation	75,000 (paid)	
Publishing Company	Tucker Publishing Group	
Publisher/Editor	Kristen Tucker	ktucker@evansvilleliving.com
Advertising Manager	Prudence Hoesli	prudence@evansvilleliving.com

See **Go Team Magazine** (listed under Sports/Specialty Publications).

See **Message** (listed under Religion/Specialty Publications).

News 4U

Mailing Address	P.O. Box 14131, Evansville IN 47728
Street Address	4 Chestnut St., Evansville IN 47713
Telephone	812-429-3907
Telephone (toll-free)	866-963-9748
Fax	812-429-3908
E-mail (news)	editor@news-4u.com
Web Site	www.news4uonline.com
Publication Date	Monthly
Circulation	25,000 (free/newsstand)
Publishing Company	Publications of News4u, Inc.
Publisher	Bashar Hamami bashar@news-4u.com
General Manager	Sharon Tindle sharon@news-4u.com
Managing Editor	Dylan Gibbs editor@news-4u.com
Advertising Manager	Lori Martin lori@news-4u.com
Production Supervisor	Amanda Smith art@news-4u.com
Notes	Features entertainment/dining/shopping news. Publishes edition for Owensboro, KY. Also publishes **Tri-State Bride** and **What's Cookin'! Magazine** (annual publications).

Our Times

Mailing Address	P.O. Box 2164, Evansville IN 47728
Street Address	605 S. Evans Ave., Evansville IN 47713
Telephone	812-426-7993
Fax	812-425-0066
E-mail	ourtimesnewspaper@juno.com
Publication Date	Bi-weekly (Friday)
Circulation	2,500 (paid)
Publishing Company	SLM Marketing Communications
Publisher/Editor	Sondra Matthews
Advertising Manager	DeMarco Hampton
Notes	Serves African-American community.

See **The Shield** (listed under College Campus/Specialty Publications).

See **K-Love** (listed under National Radio Stations).

WABX (107.5 FM) — WEJK (107.1 FM) — WIKY (104.1 FM) — WLFW (93.5 FM) WSTO (96.1 FM)

Mailing Address	P.O. Box 3848, Evansville IN 47736
Street Address	1162 Mount Auburn Rd., Evansville IN 47720
Telephone	812-424-8284
Telephone (toll-free)	800-879-3172
Telephone (news)	812-463-7985
Fax	812-426-7928
Fax (news)	812-422-1826
Fax (sales)	812-421-3273
E-mail (news)	news@wiky.com
On-air Hours	24/7
Broadcast Company	South Central Media
General Manager	Tim Huelsing	tim@southcentralmedia.com
News Director	Randy Wheeler	rwheeler@southcentralmedia.com
Accepts PSAs?	yes (contact Linda Goebel, lgoebel@southcentralmedia.com)

WABX

Web Site	www.wabx.net	
Wattage	6,000	
Format	Classic Rock	
Program Director	Rusty James	rjames@southcentralmedia.com
Sales Manager	Jaleigh Burger	jlong@southcentralmedia.com
Promotions Director	Falen Bonsett	fbonsett@southcentralmedia.com

WEJK

Web Site	www.1071jackfm.com	
Wattage	6,000	
Format	Jack-FM	
Program Director	Mark Elliott	melliott@southcentralmedia.com
Sales Manager	Paul Brayfield	pbrayfield@southcentralmedia.com
Promotions Director	James Ashley	jashley@southcentralmedia.com
Notes	Does not broadcast local news. Accepts only pre-recorded PSAs.	

WIKY

Web Site	www.wiky.com	
Wattage	50,000	
Format	Adult Contemporary	
Network	AP	
Program Director	Mark Elliott	melliott@southcentralmedia.com
Sales Manager	Paul Brayfield	pbrayfield@southcentralmedia.com
Promotions Director	James Ashley	jashley@southcentralmedia.com

WLFW

Web Site	www.935thewolf.com	
Wattage	6,000	
Format	Country	
Program Director	Rusty James	rjames@southcentralmedia.com
Sales Manager	Paul Brayfield	pbrayfield@southcentralmedia.com
Promotions Director	James Ashley	jashley@southcentralmedia.com

WSTO

Web Site	www.hot96.com	
Wattage	100,000	
Format	Contemporary Hit Radio	
Program Director	Jason Addams	jadams@southcentralmedia.com
Sales Manager	Jaleigh Long	jlong@southcentralmedia.com
Promotions Director	Falen Bonsett	fbonsett@southcentralmedia.com

WBGW (101.5 FM) — WBHW (88.7 FM)

Mailing Address	P.O. Box 4164, Evansville IN 47724
Street Address	4463 E. 1200 S., Haubstadt IN 47639
Telephone	812-386-3342
Telephone (toll-free)	800-264-5550
Fax	812-768-5552
E-mail	mail@thyword.org
Web Site	www.thyword.org
Wattage	6,000
Format	Christian Inspirational
Network Affiliations	Moody, Thy Word Network
On-air Hours	24/7
Broadcast Company	Music Ministries, Inc.
General Manager	Floyd Turner floyd.turner@thyword.org
Promotions Director	Susan Turner susan.turner@thyword.org
Accepts PSAs?	yes (contact Susan Turner)
Notes	WBGW & WBHW are simulcast. WBHW serves Loogootee. Broadcasts as WBJW FM (81.7) in Albion, Illinois. Repeats on 106.5 FM in Owensboro, KY. Stations do not broadcast local news.

WCFY (102.7 FM)

Address	4100 Millersburg Rd., Evansville IN 47725
Telephone	812-867-6464
Fax	812-867-4080
E-mail	info@99thebridge.org
E-mail (news)	happenings@99thebridge.org
Web Site	www.99thebridge.org
Wattage	100
Format	Contemporary Christian
On-air Hours	24/7
Owner	Christian Fellowship Church
General Manager	Steve Gubbins steve.gubbins@99thebridge.org
Accepts PSAs?	yes (contact happenings@99thebridge.org)
Notes	Does not broadcast local news.

WDKS (106.1 FM) — WGBF (1280 AM) — WGBF (103.1 FM) — WJLT (105.3 FM) WKDQ (99.5 FM)

Address	117 S. E. 5th St., Evansville IN 47708
Telephone	812-425-4226
Fax	812-421-0005
On-air Hours	24/7
Broadcast Company	Regent Communications
General Manager	Mark Thomas mthomas@regentcomm.com
Accepts PSAs?	yes (contact John Story, john.story@regentcomm.com
Notes	No in-house news department; broadcasts news from WEHT TV.

WDKS

Web Site	www.1061evansville.com
Wattage	6,000
Format	Adult Contemporary Hit Radio
Program/Promotions	Ryan Lewis ryano@1061evansville.com
Sales Manager	Angie Ross aross@regentcomm.com

WGBF AM

Web Site	www.newstalk1280.com
Wattage	5,000 (day); 1,000 (night)
Format	News/Talk
Networks	Citadel, Fox News, Premiere, Talk Radio Network, Westwood One
Program Director	John Story john.story@regentcomm.com
Sales Manager	Angie Ross aross@regentcomm.com
Promotions Director	Bobby Gates bgates@regentcomm.com

WGBF FM

E-mail	sandman@103gbfrocks.com
Web Site	www.103gbfrocks.com
Format	Active Rock
Program Director	Mike Sanders msanders@regentcomm.com
Sales Manager	Angie Ross aross@regentcomm.com
Promotions Director	Bobby Gates bgates@regentcomm.com
Accepts PSAs?	yes (contact Bobby Gates)

WJLT

Web Site	www.superhits1053.com
Wattage	50,000
Format	Oldies
Program/Promotions	Johnny Kincaid johnny.kincaid@regentcomm.com
Sales Manager	LaDonne Craig ladonne@wkdq.com

WKDQ

Web Site	www.wkdq.com
Wattage	100,000
Format	Country
Program Director	Jon Prell jonp@wkdq.com
Sales Manager	LaDonne Craig ladonne@wkdq.com
Promotions Director	Eric Cornish eric.cornish@regentcomm.com

WEOA (1400 AM)

Address	915 Main St., STE 001, Evansville IN 47708
Telephone	812-424-8864
Telephone (toll-free)	888-846-7867
Fax	812-424-9946
E-mail	weoa_1@yahoo.com
Wattage	1,000
Format	Urban Adult Contemporary
Network Affiliations	ABC
On-air Hours	24/7
Broadcast Company	BLS Entertainment
General Manager	Ed Lander
News Director	Larry Schweizer
Public Affairs Director	Regina Lander
Accepts PSAs?	yes (contact Regina Lander)

WGAB (1180 AM)

Mailing Address	P.O. Box 2463, Evansville IN 47728
Street Address	2324 Culverson Ave., Evansville IN 47714
Telephone	812-479-5342
Telephone (toll-free)	800-600-7230
Fax	812-474-0483
E-mail	sales@faithmusicmissions.org
Web Site	www.faith1180.com
Wattage	675
Format	Religious
On-air Hours	daytime
Broadcast Company	Faith Broadcasting
General Manager	Gayle Russ
Accepts PSAs?	yes

WNIN (88.3 FM)

Address	405 Carpenter St., Evansville IN 47708	
Telephone	812-423-2973	
Fax	812-428-7548	
Web Site	www.wnin.org	
Wattage	15,500	
Format	NPR News & Great Music	
Network Affiliations	NPR, PRI	
On-air Hours	24/7	
Broadcast Company	Tri-State Public Media	
General Manager	David Dial	ddial@wnin.org
V. P. Radio Manager	Steve Burger	sburger@wnin.org
V. P. Development	Suzanne Hudson-Smith	shudsonsmith@wnin.org
News Director	Micah Schweizer	mschweizer@wnin.org
Program Director	Daniel Moore	dmoore@wnin.org
Accepts PSAs?	yes (contact Daniel Moore)	
Notes	Non-commercial station.	

WPSR (90.7 FM)

Address	1901 Lynch Rd., Evansville IN 47711
Telephone	812-435-8241
Fax	812-435-8828
E-mail	wpsr@evsc.k12.in.us
Wattage	14,000
Format	Variety/80's Hits
Network Affiliations	Network Indiana
On-air Hours	24/7
Owner	Evansville Vanderburgh School Corp.
General Manager	Michael Reininga
Program/Public Affairs Dir.	Ruth Randall
Notes	Non-commercial station.

See **WSJD** (Princeton).

WSWI (820 AM)

Address	8600 University Blvd., Evansville IN 47712
Telephone	812-465-1665
Telephone (news)	812-464-1927
E-mail	wswi@usi.edu
Web Site	www.820theedge.com
Wattage	250
Format	Alternative
On-air Hours	daytime
Owner	University of Southern Indiana
General Manager	John Morris jmmorris@usi.edu
Accepts PSAs?	yes
Notes	Non-commercial station.

WUEV (91.5 FM)

Address	1800 Lincoln Ave., Evansville IN 47722
Telephone	812-488-2022
Fax	812-488-2320
E-mail	wuevfm@evansville.edu
Web Site	http://wuev.evansville.edu
Wattage	6,100
Format	Jazz/Sports/Vairety
On-air Hours	24/7
Owner	University of Evansville
General Manager	Brandon Gaudin
Accepts PSAs?	yes (contact Brandon Gaudin)
Notes	Non-commercial station.

WVHI (1330 AM)

Mailing Address	P.O. Box 3636, Evansville IN 47735
Street Address	114 N. W. Martin Luther King Blvd., Evansville IN 47708
Telephone	812-425-2221
Fax	812-425-2078
E-mail	krista@wvhi.com
Web Site	www.wvhi.com
Wattage	5,000
Format	Adult Contemporary Christian
Broadcast Company	Word Broadcasting
General Mgr./News Dir.	Krista Denton krista@wvhi.com
Accepts PSAs?	yes

WYIR (96.9 FM)

Mailing Address	P.O. Box 994, Newburgh IN 47529
Street Address	111 N.W. 4th St., Evansville IN 47711
Telephone	812-618-1938
E-mail	info@wyir.com
Web Site	www.wyir.com
Wattage	58
Format	Alternative/New Rock
On-air Hours	24/7
Owner	Youth Incorporated of Southern Indiana
General Manager	Jesse Fuller	jesse@youthinc.net
Executive Director	Andy Fuller	andy@wyir.com
Engineer	Skip Spence	sspence@oak.edu
Accepts PSAs?	yes

See **WYFX** (Vincennes).

WAZE TV (Channel 20)

Address	2540 Waterbridge Way, Evansville IN 47710
Telephone	812-425-1900
Fax	812-423-3405
Network Affiliation	CW
On-air Hours	24/7
Broadcast Company	Roberts Brothers Broadcasting Evansville, LLC
General Mgr./Sales Mgr.	Greg Pittman	gregp@roberts-companies.com
Creative Services Director	Jim Alexander	jima@roberts-companies.com
Accepts PSAs?	yes (contact Jim Alexander)
Notes	Does not broadcast local news.

WEHT TV (Channels 25.1/ABC & 25.2/Retro Television Network)

Mailing Address	P.O. Box 25, Evansville IN 47701
Street Address	800 Marywood Dr., Henderson KY 42420
Telephone	812-424-9215
Telephone (toll-free)	800-879-8542
Telephone (news)	800-879-8549
Fax	270-826-6823
Fax (news)	270-827-0561
E-mail (news)	news@news25.us
Web Site	www.news25.us
Network Affiliation	ABC
On-air Hours	24/7
Broadcast Company	Gilmore Broadcasting Corp.
General Manager	Doug Padgett	dpadgett@news25.us
News Director	Mark Glover	mglover@news25.us
News Assignment Editor	Lewis Chaney	lchaney@news25.us
Sports Director	Lance Wilkerson	lwilkerson@news25.us
General Sales Manager	Mike Riley	mriley@news25.us
Program/Public Affairs Dir.	Ginny Powers	gpowers@news25.us
Promotions Director	Melisse Marks	mmarks@news25.us
Accepts PSAs?	yes (contact Cathy White, cwhite@news25.us)

WEVV TV (Channels 44.1/CBS & 44.2/MyNetwork TV)

Address	44 Main St., Evansville IN 47708
Telephone	812-464-4444
Fax	812-465-4559
E-mail	info@wevv.com
Web Site	www.wevv.com
Network Affiliations	CBS & MyNetwork TV
On-air Hours	24/7
Broadcast Company	Communications Corporation of Indiana
General Manager	Tim Black	tim.black@wevv.com
General Sales Manager	Greg Murdach	greg.murdach@wevv.com
Local Sales Manager	Brian Tornatta	brian.tornatta@wevv.com
Program Director	Joanne Provenzano	joanne.provenzano@wevv.com
Promotions Director	Eric Stremming	eric.stremming@wevv.com
Accepts PSAs?	yes (contact Eric Stremming)
Notes	Does not broadcast local news.

WFIE TV (Channels 46.1/NBC; 46.2/24-7 Weather Now; & 46.3/This TV)

Mailing Address	P.O. Box 1414, Evansville IN 47701
Street Address	1115 Mount Auburn Rd., Evansville IN 47720
Telephone	812-426-1414
Telephone (toll-free)	800-832-0014
Telephone (news)	812-425-3026
Fax	812-426-1945
Fax (news)	812-428-2228
Fax (sales)	812-425-2482
E-mail (news)	sgalloway@14wfie.com
Web Site	www.14wfie.com
Network Affiliation	NBC
On-air Hours	24/7
Broadcast Company	Raycom Media
General Manager	Debbie Bush	dbush@14wfie.com
News Director	C. J. Hoyt	cjhoyt@14wfie.com
News Assignment Editor	Scott Galloway	sgalloway@14wfie.com
Sports Director	Mike Blake	mblake@14wfie.com
General Sales Manager	Laura Lovejoy	llovejoy@14wfie.com
Program Director	Kirk Williams	kwilliams@14wfie.com
Public Affairs/Promotions	Adam Frary	afrary@14wfie.com
Accepts PSAs?	yes (contact Adam Frary)

WNIN TV (Channels 9.1/PBS, 9.2/WNIN Local; & 9.3/WNIN-FM radio simulcast)

Address	405 Carpenter St., Evansville IN 47708
Telephone	812-423-2973
Telephone (toll-free)	800-423-5678
Fax	812-428-7548
Web Site	www.wnin.org
Network Affiliation	PBS
On-air Hours	24/7
Broadcast Company	Tri-State Public Media Inc.
General Manager	David Dial	ddial@wnin.org
General Sales Manager	Tonya Wolf	twolf@wnin.org
Program Director	Bonnie Rheinhardt	brheinhardt@wnin.org
Chief Engineer	Don Hollingsworth	dhollingsworth@wnin.org
Accepts PSAs?	yes (contact Tony Voss, tvoss@wnin.org)
Notes	Non-commercial station.

WTSN TV (Channel 36)

Address	300 S.E. Riverside Dr., STE 100, Evansville IN 47713
Telephone	812-759-8191
Fax	812-759-0235
E-mail	info@wtsn36.com
Web Site	www.wtsn36.com
Network Affiliation	America One
On-air Hours	24/7
Traffic Manager	Jennifer LeRoy
Accepts PSAs?	yes (submit DVD or MPEG)

WTVW TV (Channel 28)

Address	477 Carpenter St., Evansville IN 47708	
Telephone	812-424-7777	
Telephone (toll-free)	800-511-6009	
Telephone (news)	812-421-4030	
Fax	812-421-4040	
Fax (news)	812-421-7289	
E-mail (news)	newstips@wtvw.com	
Web Site	www.tristatehomepage.com	
Network Affiliation	Fox	
On-air Hours	24/7	
Broadcast Company	Nexstar Broadcasting	
General Manager	Mike Smith	msmith@wtvw.com
News Director	Bob Walters	bwalters@wtvw.com
News Assignment Editor	Warren Korff	wkorff@wtvw.com
Sports Director	Doug Kufner	dkufner@wtvw.com
General Sales Manager	Jeff Fisher	jfisher@wtvw.com
Local Sales Manager	Jay Hiett	jayh@wtvw.com
Promotions Director	Rob Underwood	runderwood@wtvw.com
Production Director	John Payne	jpayne@wtvw.com
Accepts PSAs?	yes (contact Rob Underwood)	

FAIRMOUNT

Grant County

News-Sun — Riverside Palaver

Mailing Address	P.O. Box 25, Fairmount IN 46928
Street Address	122 S. Main, Fairmount IN 46928
Telephone/Fax	765-948-4165
Publication Date	Weekly: Wednesday (News-Sun)
	Weekly: Thursday (Riverside Palaver)
Circulation	4,400-free/mailed (News-Sun)
	4,000-free/mailed (Riverside Palaver)
Publishing Company	Allen Terhune & Associates
Publisher	Jim Terhune jtn_s@frontiernet.net
Notes	Riverside Palaver serves Gas City and Jonesboro.

FERDINAND

Dubois County

Ferdinand News

Mailing Address	P.O. Box 38, Ferdinand IN 47532
Street Address	113 W. 6th St., Ferdinand IN 47532
Telephone	812-367-2041
Fax	812-367-2371
E-mail	thenews@psci.net
E-mail (advertising)	ads@psci.net
Publication Date	Weekly (Wednesday)
Circulation	3,100 (paid)
Publishing Company	Dubois-Spencer Counties Publishing
Publishers	Paul & Miriam Ash; Richard & Kathy Tretter
Editor/Advertising Mgr.	Kathy Tretter	ferdnews@psci.net
Sports Editor	Brian Bohne	dssports@psci.net

FISHERS

Hamilton County

See **atGeist Community Newsletter** (Indianapolis).

Indianapolis Star—North Bureau

Address	13095 Publisher's Dr., Fishers IN 46038
Telephone	317-444-4444
Web Sites	www.carmelstar.com
	www.fishersstar.com
	www.noblesville star.com
	www.northindystar.com
	www.westfieldstar.com
	www.zionsvillestar.com
Publication Date	Weekly (Thursday)
Publishing Company	Gannett Co. Inc.
Suburban Editor	Kevin Morgan	kevin.morgan@indystar.com
Notes	Main office in Indianapolis. Publishes community editions for Carmel, Fishers, Noblesville & Westfield (Hamilton County); North Indy (Marion County); and Zionsville (Boone County).

FLAT ROCK

Shelby County

See **WJCF** (Greenfield).

FLORA

Carroll County Comet
Carroll County

Mailing Address	P.O. Box 26, Flora IN 46929
Street Address	14 E. Main St., Flora IN 46929
Telephone	574-967-4135
Fax	574-967-3384
E-mail (news)	editor@carrollcountycomet.com
Web Site	www.carrollcountycomet.com
Publication Date	Weekly (Wednesday)
Circulation	4,800 (paid)
Publishing Company	Carroll Papers Inc.
Publisher/Editor	Susan Scholl	editor@carrollcountycomet.com
Publisher/Advertising Mgr.	Joe Moss	comet@carrollcountycomet.com
Notes	Bureau in Delphi.

FORT BRANCH

South Gibson Star-Times
Gibson County

Mailing Address	P.O. Box 70, Fort Branch IN 47648
Street Address	203 S. McCreary St., Fort Branch IN 47648
Telephone	812-753-3553
Fax	812-753-4251
E-mail	news@sgstartimes.com
Publication Date	Weekly (Tuesday)
Circulation	3,000 (paid/mailed)
Publishing Company	Pike Publishing
Publisher	Frank Heuring	fheuring@blueriver.net
Editor	Jessica Alaimo	editor@sgstartimes.com
Advertising Manager	John Heuring	ads@sgstartimes.com
Sports Editor	Jim Capozella	sports@sgstartimes.com
TMC/Shopper	The Bulletin (weekly)

See **WENS** (Greenfield).

FORT WAYNE

Aboite & About — Dupont Valley Times
East Allen County Times — George Town Times — St. Joe Times
Allen County

Mailing Address	P.O. Box 11448, Fort Wayne IN 46858
Street Address	826 Ewing St., Fort Wayne IN 46802
Telephone	260-426-5511
Fax	260-426-1551
E-mail	pr@timespubs.com
Web Site	www.timespubs.com
Publication Date	Monthly
Circulation	80,000-free/mailed (combined)
Publishing Company	Times Community Publications (division of KPC Media Group)
General Manager	Lynn Sroufe	lsroufe@kpcnews.net
Editor	Sue Reeves	pr@timespubs.com
Advertising Manager	Sherri Ayres	sayres@kpcnews.net
Notes	East Allen County Times serves New Haven.

See **Business People** (listed under Business/Specialty Publications).

See **Communicator** (listed under College Campus/Specialty Publications).

El Mexicano Newspaper

Address	2301 Fairfield Ave., STE 102, Fort Wayne IN 46807
Telephone	260-456-6843
Telephone (toll-free)	877-505-9435
Telephone (news)	260-704-0682
Fax	260-456-2535
E-mail	elmexica@earthlink.net
Web Site	www.elmexicanonews.com
Publication Date	Monthly
Circulation	10,000 (free/delivered & mailed)
Publisher/Editor	Fernando Zapari
Notes	Serves Hispanic community.

Fort Wayne Living

Address	7729 Westfield Dr., Fort Wayne IN 46825	
Telephone	260-497-0433	
Fax	260-497-0822	
E-mail	jcopeland@businesspeople.com	
E-mail (news)	editor@businesspeople.com	
Web Site	www.businesspeople.com	
Publication Date	Quarterly	
Circulation	21,000 (paid)	
Publishing Company	Michiana Business Publications Inc.	
Publisher	Daniel C. Copeland	dcopeland@businesspeople.com
Editor	Amber Recker	arecker@businesspeople.com

Fort Wayne Monthly Magazine

Address	600 W. Main St., Fort Wayne IN 46802	
Telephone (toll-free)	800-444-3303	
Fax	270-461-8863	
E-mail	editor@fwn.fortwayne.com	
Web Site	www.ftwaynemagazine.com	
Publication Date	Monthly	
Circulation	12,500 (paid)	
Publishing Company	Fort Wayne Newspapers	
Publisher	Lisa Goodman	lgoodman@fwn.fortwayne.com
Editor	Connie Haas Zuber	chaaszuber@fwn.fortwayne.com

Fort Wayne Reader — Ink Newspaper

Address	1301 Lafayette St., STE 202, Fort Wayne IN 46802
Fax	260-420-3210
Publishing Company	Diversity Media Group

Fort Wayne Reader

Telephone	260-420-8580
E-mail	mikes@fortwaynereader.com
Web Site	www.fortwaynereader.com
Publication Date	Semi-monthly (1st & 3rd Friday)
Circulation	10,000 (free/newsstand)
Publisher/Editor	Michael Summers mikes@fortwaynereader.com
Notes	Arts & entertainment publication.

Ink Newspaper

Telephone	260-420-3200
E-mail	editor@inknewsonline.com
Web Site	www.inknewsonline.com
Publication Date	Weekly (Friday)
Circulation	9,500 (paid)
Publisher/Ad Mgr.	Terri Miller
Editor	Vince Robinson
Notes	Serves African-American community.

Frost Illustrated

Address	3121 S. Calhoun St., Fort Wayne IN 46807
Telephone	260-745-0552
Fax	260-745-9503
E-mail	frostnews@aol.com
Web Site	www.frostillustrated.com
Publication Date	Weekly (Wednesday)
Circulation	8,500 (paid)
Publishing Company	Frost Inc.
Publisher	Edward N. Smith Sr.
Managing Editor	Michael Patterson
Advertising Manager	Edna M. Smith frostads@aol.com
Notes	Serves African-American community.

See **Greater Fort Wayne Business Weekly** (listed under Business/Specialty Publications).

See **Greater Fort Wayne Family** (listed under Parenting/Specialty Publications).

Journal Gazette

Mailing Address	P.O. Box 88, Fort Wayne IN 46801
Street Address	600 W. Main St., Fort Wayne IN 46802
Telephone	260-461-8444
Telephone (toll-free)	800-444-3303
Telephone (news)	260-461-8428
Fax	260-461-8648
Fax (news)	260-461-8893
Fax (advertising)	260-461-8230
E-mail (news)	jgnews@jg.net
Web Site	www.journalgazette.net
Publication Date	Daily (Sunday-Saturday)
Circulation	60,000-paid (daily); 118,000-paid (Sunday)
Publishing Company	Journal Gazette Co.
Publisher	Julie Inskeep	jinskeep@jg.net
Editor	Craig Klugman	cklugman@jg.net
Managing Editor	Sherry Skufca	sskufca@jg.net
Advertising Manager	Henry Phillips	hphillips@fwn.fortwayne.com
Sports Editor	Mark Jaworski	mjaworski@jg.net
Metro Editor	Tom Germuska	tgermuska@jg.net
Business Editor	Lisa Green	lgreen@jg.net
Editorial Page Editor	Tracy Warner	twarner@jg.net
Notes	Bureaus in Indianapolis (see listing) and Washington D.C.: 529 14th St. N.W., STE 551, Washington DC 20045; 202-879-6710 (phone); 202-879-6712 (fax); Sylvia Smith, Bureau Chief; sylviasmith@jg.net). The Journal Gazette and the News Sentinel have separate editorial staffs but have joint finance/advertising/production operations under the name Fort Wayne Newspapers.

News-Sentinel

Mailing Address	P.O. Box 102, Fort Wayne IN 46801
Street Address	600 W. Main St., Fort Wayne IN 46802
Telephone	260-461-8439
Telephone (toll-free)	800-444-3303
Telephone (news)	260-461-8354
Fax (news)	260-461-8817
E-mail (news)	nsmetro@news-sentinel.com
Web Site	www.news-sentinel.com
Publication Date	Daily (Monday-Saturday)
Circulation	25,000 (paid)
Publishing Company	Ogden Newspapers
Publisher	Michael Christman	mchristman@fwn.fortwayne.com
Editor	Kerry Hubartt	khubartt@news-sentinel.com
Managing Editor	Mary Lou Brink	mbrink@news-sentinel.com
Advertising Manager	Henry Phillips	hphillips@fwn.fortwayne.com
Sports Editor	Tom Davis	tdavis@news-sentinel.com
News/Presentation Editor	Jon Swerens	jswerens@news-sentinel.com
Metro Editor	Elbert Starks III	estarks@news-sentinel.com
Editorial Page Editor	Leo Morris	lmorris@news-sentinel.com
Notes	The Journal Gazette and the News Sentinel have separate editorial staffs but have joint finance/advertising/production operations under the name Fort Wayne Newspapers.

See **Today's Catholic** (listed under Religion/Specialty Publications).

Waynedale News

Address	2700 Lower Huntington Rd., Fort Wayne IN 46809
Telephone	260-747-4535
Fax	260-747-5529
E-mail (news)	news@waynedalenews.com
Web Site	www.waynedalenews.com
Publication Date	Bi-weekly
Circulation	10,700 (free/delivered & mailed)
Owners/Publishers	Alex Cornwell & Michael Albercio
Editor	Cindy Cornwell
Advertising Manager	Alex Cornwell
Notes	Serves greater Waynedale area.

Whatzup

Address	1747 St. Mary's Ave., Fort Wayne IN 46808
Telephone	260-424-4200
Fax	260-424-6600
E-mail	info.whatzup@gmail.com
Web Site	www.whatzup.com
Publication Date	weekly (Thursday)
Circulation	12,600 (free)
Publishing Company	Ad Media
Notes	Entertainment publication.

WAJI (95.1 FM) — WLDE (101.7 FM)

Address	347 W. Berry St., STE 600, Fort Wayne IN 46802	
Telephone	260-423-3676	
Fax	260-422-5266	
On-air Hours	24/7	
Broadcast Company	Sarkes Tarzian Inc.	
General Manager	Lee Tobin	ltobin@stfortwayne.com
Sales Manager	Shelly Steckler	ssteckler@stfortwayne.com

WAJI

E-mail (news)	Jeannette@waji.com	
Web Site	www.waji.com	
Wattage	39,000	
Format	Adult Contemporary	
News/Public Affairs	Jeannette Rinard	jeannette@waji.com
Program Director	Barb Richards	barbrichards@waji.com
Sports Director	Dirk Rowley	dirk@waji.com
Promotions Director	Marti Taylor	mtaylor@waji.com
Accepts PSAs?	yes (contact Jeannette Rinard)	

WLDE

E-mail (news)	carriebud@wlde.com	
Web Site	www.wlde.com	
Wattage	3,000	
Format	Classic Hits	
News/Public Affairs	Carrie Wellman	carriebud@wlde.com
Program Director	Chris Didier	captainchris@wlde.com
Sports Director	Jim Reed	greekdaddy@wlde.com
Promotions Director	Katrina Newman	knewman@wlde.com
Accepts PSAs?	yes (contact Chris Didier)	

WBCL (90.3 FM)

Address	1025 W. Rudisill Blvd., Fort Wayne IN 46807
Telephone	260-745-0576
Telephone (news)	260-745-0578
Fax	260-456-2913
Fax (news)	260-745-2001
E-mail	wbcl@wbcl.org
Web Site	www.wbcl.org
Wattage	50,000
Format	Contemporary Christian
On-air Hours	24/7
Owner	Taylor University
General Manager	Marsha Bunker
News Director	Larry Bower
Program Director	Scott Tsuleff
Promotions Director	Jill Johnston
Accepts PSAs?	yes (contact Ron Schneemann)
Notes	Non-commercial station. Repeats on 106.1 FM (Muncie) and 97.7 FM (Adrian, MI). Simulcasts as WBCY (89.5 FM) serving Northwest Ohio; WBCJ (88.1 FM) serving North Central Ohio: and WCVM (94.7 FM) serving South Central Michigan.

WBNI (94.1 FM) — WBOI (89.1FM) — WCKZ (91.3 FM)

Mailing Address	P.O. Box 8459, Fort Wayne IN 46898	
Street Address	3204 Clairmont Ct., Fort Wayne IN 46808	
Telephone	260-452-1189	
Telephone (toll-free)	800-471-9264	
Fax	260-452-1188	
Web Site	www.nipr.fm	
On-air Hours	24/7	
Broadcast Company	Northeast Indiana Public Radio	
General Mgr./Sales Mgr.	Joan Brown	jbrown@nipr.fm
Music Director	Janice Furtner	jfurtner@nipr.fm
Chief Engineer	Ed Didier	edidier@nipr.fm
Accepts PSAs?	yes-via e-mail (contact Joan Brown)	
Notes	Non-commercial stations.	

WBNI & WCKZ

Wattage	2,000
Format	Classical Music
Networks	APM, NPR
Notes	WBNI & WCKZ are simulcast. Stations do not broadcast local news. WCKZ serves Orland and repeats on 88.7 FM (Fort Wayne).

WBOI

E-mail (news)	pshaull@nipr.fm	
Wattage	50,000	
Format	News/Jazz	
Network	NPR	
News Director	Phil Shaull	pshaull@nipr.fm

WBTU (93.3 FM) — WJFX (107.9 FM) — WJOE (106.3 FM)

Address	2100 Goshen Rd., STE 232, Fort Wayne IN 46808
Telephone	260-482-9288
Fax	260-482-8655
On-air Hours	24/7
Broadcast Company	Oasis Radio Group
Market Manager	Pete DeSimone — pete.desimone@oasisradiogroup.com
Sales Manager	Bart Schacht
Public Affairs Director	Teri Armstrong — teri.armstrong@oasisradiogroup.com
Promotions Director	Lynn Williams — lynn.williams@oasisradiogroup.com
Accepts PSAs?	yes (contact Teri Armstrong)

WBTU
Web Site	www.us933.us
Wattage	50,000
Format	Country
Program Director	Dave Steele — dave.steele@oasisradiogroup.com

WJFX
Web Site	www.hot1079online.com
Wattage	6,000
Format	Contemporary Hit Radio
Program Director	Phil Becker — phil.becker@oasisradiogroup.com

WJOE
Web Site	www.1063joefm.com
Wattage	6,000
Format	Adult Hits
Program Director	Phil Becker — phil.becker@oasisradiogroup.com

WBYR (98.9 FM) — WFWI (92.3 FM)

Address	1005 Production Rd., Fort Wayne IN 46808
Telephone	260-471-5100
Telephone (toll-free)	800-432-2327
Telephone (news)	260-447-6397
Fax	260-471-5224
Fax (news)	260-447-7546
E-mail	spots@wbyr.com
On-air Hours	24/7
Broadcast Company	Pathfinder Communications
General Manager	Jim Allgeier — jallgeier@federatedmedia.com
Promotions Director	Jenna Tucker — jtucker@federatedmedia.com

WBYR
Web Site	www.989thebear.com
Wattage	50,000
Format	Active Rock
Program Director	Stiller — stiller@989thebear.com
Sales Manager	Ann Tenney — anntenney@989thebear.com

WFWI
Web Site	www.923thefort.com
Wattage	4,000
Format	Classic Rock
Program Director	Billy Elvis — belvis@923thefort.com
Sales Manager	Suzee Leavell — sleavell@federatedmedia.com

WBZQ (1300 AM) — **WMYQ** (101.1 FM)

Mailing Address	P.O. Box 5570, Fort Wayne IN 46895
Street Address	3402-4 N. Anthony Blvd., Fort Wayne IN 46805
Telephone	260-482-8500
E-mail	q101@myq101.com
E-mail (news)	news@myq101.com
Web Site	www.myq101.com
Wattage	500 (WBZQ)
	6,000 (WMYQ)
Format	Variety/Oldies (WBZQ)
	Classic Hits (WMYQ)
Network Affiliations	ABC, Chicago Cubs, White Sox
Broadcast Company	Larko Communications
General Mgr./Sales Mgr.	Chris Larko chrislarko@myq101.com
News Director	Mike Nelson news@myq101.com
Accepts PSAs?	yes (contact communitycalendar@myq101.com)
Notes	Prefers news releases via e-mail.
	WBZQ serves Huntington/Fort Wayne. WMYQ serves Warsaw/Fort Wayne.

WCYT (91.1 FM)

Address	4310 Homestead Rd., Fort Wayne IN 46814
Telephone	260-431-2271
Fax	260-431-2299
E-mail	comment@wcyt.org
Web Site	www.wcyt.org
Wattage	120
Format	Alternative
On-air Hours	24/7
Owner	Southwest Allen County Schools
General Manager	Adam Schenkel adam@wcyt.org
Accepts PSAs?	yes (contact Adam Schenkel)
Notes	Non-commercial station.

WFCV (1090 AM)

Address	3737 Lake Ave., Fort Wayne IN 46805
Telephone	260-423-2337
Fax	260-423-6355
E-mail	wfcv@bottradionetwork.com
Web Site	www.bottradionetwork.com
Wattage	2,500
Format	Christian News & Information/All Talk
Network Affiliations	Bott Radio Network
On-air Hours	daytime
Broadcast Company	Bott Radio Network
General Mgr./Sales Mgr.	Dale Gerke dgerke@bottradionetwork.com
Public Affairs/Promotions	Renee Bodine rbodine@bottradionetwork.com
Operations Manager	Kathy McClish kmcclish@bottradionetwork.com
Accepts PSAs?	yes (contact Renee Bodine)

WGBJ (102.3 FM)

Address	4534 Parnell Ave., Fort Wayne IN 46825
Telephone	260-482-4444
Fax	260-482-4410
Web Site	www.thenewpower1023.com
Wattage	6,000
Format	Contemporary Top 40
On-air Hours	24/7
Broadcast Company	Fort Wayne Broadcasting
General Manager	Alex Archer aarcher@thenewpower1023.com
Promotions Director	Vincent Wilson vwilson@thenewpower1023.com
Accepts PSAs?	yes (contact Angie Phillips, aphillips@thenewpower1023.com)

WGL (1250 AM) — WGL (102.9 FM) — WNHT (96.3 FM) — WXKE (103.9 FM)

Address	2000 Lower Huntington Rd., Fort Wayne IN 46819
Telephone	260-747-1511
Fax	260-747-3999
E-mail (news)	stone@summitcityradio.com
On-air Hours	24/7
Broadcast Company	Summit City Radio Group
News/Public Affairs Dir.	Leslie Stone stone@summitcityradio.com
Sales Manager	Dave Reithmiller dave@summitcityradio.com
Promotions Director	Michele Tessier mtessier@summitcityradio.com
Operations Manager	J. J. Fabini jj@summitcityradio.com
Accepts PSAs?	yes (contact Leslie Stone)

WGL AM

Web Site	www.1250theriver.com
Wattage	2,500
Format	Adult Standards
Networks	CBS, Dial Global
Program Director	J. J. Fabini jj@summitcityradio.com

WGL FM

Web Site	www.1029theriver.com
Wattage	2,500
Format	Adult Standards
Networks	CBS, Dial Global, Hoosier Ag
Program Director	J. J. Fabini jj@summitcityradio.com

WNHT

Web Site	www.wild963.com
Wattage	6,700
Format	Hip Hop/R&B
Network	CBS
Program Director	Philip Spencer shady@summitcityradio.com

WXKE

Web Site	www.rock104radio.com
Wattage	3,000
Format	Classic Rock
Program Director	Doc West doc@summitcityradio.com

WKJG (1380 AM) — WMEE (97.3 FM) — WOWO (1190 AM) — WQHK (105.1 FM)

Mailing Address	P.O. Box 6000, Fort Wayne IN 46896	
Street Address	2915 Maples Rd., Fort Wayne IN 46816	
Telephone	260-447-5511	
Telephone (news)	260-447-6397	
Fax	260-447-7546	
E-mail (news)	aober@federatedmedia.com	
On-air Hours	24/7	
General Manager	Mark DePrez	mdeprez@federatedmedia.com
News Director	Andy Ober	aober@federatedmedia.com
Promotions Director	Brian Sheikh	bsheikh@federatedmedia.com

WKJG

Wattage	5,000	
Format	Sports	
Network	ESPN	
Broadcast Company	Federated Media Broadcasting	
Program Director	Dan Mandis	dmandis@federatedmedia.com
Sports Director	Jim Shovlin	jshovlin@wowo.com
Sales Manager	Ben Saurer	bsaurer@federatedmedia.com
Accepts PSAs?	yes (contact Andy Ober)	

WMEE

Web Site	www.wmee.com	
Wattage	50,000	
Format	Hot Adult Contemporary	
Broadcast Company	Federated Media Broadcasting	
Program Director	Rob Kelley	rkelley@federatedmedia.com
Sales Manager	Mark Osborn	mosborn@federatedmedia.com
Accepts PSAs?	yes (contact Rob Kelley)	

WOWO

Web Site	www.wowo.com	
Wattage	50,000	
Format	News/Talk	
Networks	Fox, Network Indiana	
Broadcast Company	Federated Media Broadcasting	
Program Director	Dan Mandis	dmandis@federatedmedia.com
Sports Director	Jim Shovlin	jshovlin@wowo.com
Sales Manager	Ben Saurer	bsaurer@federatedmedia.com
Accepts PSAs?	yes (contact Andy Ober)	

WQHK

Web Site	www.k105fm.com	
Wattage	25,000	
Format	Country	
Broadcast Company	Jam Communications	
Program Director	Rob Kelley	rkelley@federatedmedia.com
Sales Manager	Joel Pyle	jpyle@federatedmedia.com
Accepts PSAs?	yes (contact Rob Kelley)	

WLAB (88.3 FM)

Mailing Address	6600 N. Clinton St., Fort Wayne IN 46825
Street Address	8 Martin Luther Dr., Fort Wayne IN 46825
Telephone	260-483-8236
Telephone (toll-free)	800-359-8816
Fax	260-482-7707
Web Site	www.star883.com
Wattage	3,200
Format	Christian Adult Contemporary
On-air Hours	24/7
Broadcast Company	STAR Educational Media Network
General Manager	Melissa Montana — melissa@star883.com
Program Director	Don Buettner — don@star883.com
Promotions Director	John O'Rourke — johno@star883.com
Strategic Planning & Development	Richard Cummins — richard@star883.com
Accepts PSAs?	yes (contact psa@star883.com)
Notes	Non-commercial station. Broadcasts local news in the morning.

WLYV (1450 AM)

Address	4705 Illinois Rd., STE 104, Fort Wayne IN 46804
Telephone	260-436-9598
Telephone (toll-free)	888-436-1450
Fax	260-432-6179
E-mail	info@redeemerradio.com
Web Site	www.redeemerradio.com
Wattage	1,000
Format	Catholic
Network Affiliations	Ave Maria, EWTN
On-air Hours	24/7
Broadcast Company	Fort Wayne Catholic Radio Group
Executive Director	Dave Stevens
Program Director	Patty Becker
Sports Director	Sean McBride
Accepts PSAs?	yes (contact Patty Becker)
Notes	Does not broadcast local news.

WNUY (100.1 FM)

Address	4714 Parnell Ave., Fort Wayne IN 46825
Telephone	260-469-2412
Telephone (toll-free)	800-946-9689
Fax	260-484-3504
E-mail	ken@fm100talks.com
Web Site	www.fm100talks.com
Wattage	6,000
Format	Talk/Sports
Network Affiliations	Fox Sports, IU Sports Network, Sports USA, Westwood One, Ball State University Sports Radio Network
On-air Hours	24/7
Broadcast Company	Independence Media of Indiana
General Mgr./Sales Mgr.	Kenneth Allen — ken@fm100talks.com
Program Director	Pete LaFaucia — pete@fm100talks.com
Office Manager	Diane Current — diane@fm100talks.com
Accepts PSAs?	yes (contact Diane Current)

WQSW-LP (100.5 FM)

Address	3510 Stellhorn Rd., STE A, Fort Wayne IN 46815
Telephone	260-407-8300
Fax	260-969-5686
E-mail	stellarwomen@onecommail.com
Web Site	www.stellarwomen.com
Wattage	100
Format	Gospel
On-air Hours	24/7
Owner	Quasi Inc.
Executive Director	Deborah Godwin-Starks
Accepts PSAs?	yes (contact stellarwomen@onecommail.com)
Notes	Does not broadcast local news.

WANE TV (Channel 31)

Address	2915 W. State Blvd., Fort Wayne IN 46808	
Telephone	260-424-1515	
Telephone (news)	260-422-5644	
Fax (news)	260-424-6054	
E-mail (news)	newsrelease@wane.com	
Web Site	www.wane.com	
Network Affiliation	CBS	
On-air Hours	24/7	
Broadcast Company	LIN Television Corp.	
General Manager	Alan Riebe	alan.riebe@wane.com
News Director	Ted Linn	ted.linn@wane.com
News Assignment Editor	Scott Murray	scott.murray@wane.com
Sports Director	Glenn Marini	glenn.marini@wane.com
General Sales Manager	Tom Antisdel	tom.antisdel@wane.com
Program Director	Nancy Applegate	nancy.applegate@wane.com
Public Affairs Director	April McCampbell	april.mccampbell@wane.com
Promotions Director	Jerry Grider	jerry.grider@wane.com
Accepts PSAs?	yes (contact April McCampbell)	

WFFT TV (Channel 55)

Mailing Address	P.O. Box 8655, Fort Wayne IN 46898	
Street Address	3707 Hillegas Rd., Fort Wayne IN 46808	
Telephone	260-471-5555	
Fax	260-484-4331	
E-mail	fox55@wfft.com	
Web Site	www.fortwaynehomepage.net	
Network Affiliation	Fox	
On-air Hours	24/7	
Broadcast Company	Nexstar Broadcasting	
General Manager/ General Sales Manager	Bill Ritchhart	britchhart@wfft.com
News Director	Jim Blue	jblue@wfft.com
Local Sales Manager	Lynn Dziedzic	ldziedzic@wfft.com
Public Affairs/Promotions	John Parker	jparker@wfft.com
Accepts PSAs?	yes (contact Lis Moldes, emoldes@wfft.com)	

WFWA TV (Channels 39.1/PBS; 39.2/Kids 39; 39.3/Create TV; & 39.4/39-4 You)

Address	2501 E. Coliseum Blvd., Fort Wayne IN 46805
Telephone	260-484-8839
Fax	260-482-3632
E-mail	info@wfwa.org
Web Site	www.wfwa.org
Network Affiliation	PBS
On-air Hours	24/7
Broadcast Company	Fort Wayne Public Television
General Manager	Bruce Haines	brucehaines@wfwa.org
Program Director	Kris Hensler	krishensler@wfwa.org
Public Affairs Director	Zeke Bryant	zekebryant@wfwa.org
Promotions Director	Mark Ryan	markryan@wfwa.org
Accepts PSAs?	yes (contact Zeke Bryant)
Notes	Non-commercial station. Does not broadcast local news.

See **WFWC TV** (Auburn).

WINM TV (Channels 12 & 38)

Mailing Address	P.O. Box 159, Butler IN 46721
Street Address	3737 Lake Ave., Fort Wayne IN 46805
Telephone	260-483-9809
Fax	260-422-5024
E-mail	winm@tct.tv
Network Affiliation	TCT Ministries
On-air Hours	24/7
Broadcast Company	Tri-State Christian
General Manager	Leo Vogt
Accepts PSAs?	yes
Notes	Non-commercial station. Does not broadcast local news.
	Programming & PSAs managed at corporate office in Marion, IL.

WISE TV (Channels 18.1/NBC; 18.2/MyNetwork TV; & 18.3/INC Now-24-hour news)
WPTA TV (Channels 24.1/ABC; 24.2/The CW; & 24.3/VIPIR Channel-24-hour weather)

Mailing Address	P.O. Box 2121, Fort Wayne IN 46801
Street Address	3401 Butler Rd., Fort Wayne IN 46808
Telephone	260-483-0584
Telephone (news)	260-483-8111
Fax	260-483-1835
Fax (news)	260-484-8240
Fax (sales)	260-483-2568
E-mail	inc@incnow.tv
E-mail (news)	newsroom@incnow.tv
Web Site	www.incnow.tv
Network Affiliations	NBC & MyNetwork TV (WISE)
	ABC & The CW (WPTA)
On-air Hours	24/7
Broadcast Company	Granite Broadcasting Corp. (WISE)
	Malara Broadcasting Group (WPTA)

General Manager	Jerry Giesler	jerryg@incnow.tv
News Director	Peter Neumann	petern@incnow.tv
News Assignment Editor	Maureen Mespell	maureenm@incnow.tv
Sports Director	Dean Pantazi	deanp@incnow.tv
Promotions Director	Tad Frank	tadf@incnow.tv
Hub Operations Manager	Jim Turcovsky	jimt@incnow.tv
Director of Sales	Daniel Hoffman	danh@incnow.tv
Station Manager (WPTA)	Doug Barrow	dougb@incnow.tv
Accepts PSAs?	yes (contact Jim Turcovsky)	

FORTVILLE

Hancock County

See **Fortville-McCordsville Reporter** (Greenfield).

FOWLER

Benton County

Benton Review

Mailing Address	P.O. Box 527, Fowler IN 47944	
Street Address	204 N. Adams, Fowler IN 47944	
Telephone	765-884-1902	
Fax	765-884-8110	
E-mail	bentonreview@sbcglobal.net	
Web Site	www.thebentonreview.com	
Publication Date	Weekly (Wednesday)	
Circulation	3,000 (paid)	
Owner/Publisher/Editor	Karen Hall Moyars	bentonreview@sbcglobal.net

FRANCESVILLE

Pulaski County

Francesville Tribune

Mailing Address	P.O. Box 458, Francesville IN 47946
Street Address	111 E. Montgomery St., Francesville IN 47946
Telephone/Fax	219-567-2221
E-mail	francesville.tribune@ffni.com
Publication Date	Weekly (Thursday)
Circulation	950 (paid)
Owner/Publisher	Steve Sewell
Editor	Carla Sewell

FRANKFORT

Clinton County

Frankfort Times

Mailing Address	P.O. Box 9, Frankfort IN 46041	
Street Address	251 E. Clinton St., Frankfort IN 46041	
Telephone	765-659-4622	
Fax	765-654-7031	
E-mail (news)	news@ftimes.com	
Web Site	www.ftimes.com	
Publication Date	Daily (Monday-Saturday)	
Circulation	6,300 (paid)	
Publishing Company	Paxton Media Group	
Publisher	Terry Ward	tward@ftimes.com
Managing Editor	Brian Peloza	bpeloza@ftimes.com
Sports Editor	Phil Friend	sports@ftimes.com

WILO (1570 AM) — WSHW (99.7 FM)

Mailing Address	P.O. Box 545, Frankfort IN 46041	
Street Address	1401 W. Barner St., Frankfort IN 46041	
Telephone	765-659-3338	
Telephone (toll-free)	800-447-4463	
Fax	765-654-3484	
E-mail (news)	newsroom@kasparradio.com	
Network Affiliations	USA	
Broadcast Company	Kaspar Broadcasting Co., Inc.	
CEO	Vern Kaspar	
General Mgr./Sales Mgr.	Russ Kaspar	rk@kasparradio.com
News Director	Mike Reppert	newsroom@kasparradio.com
Program/Public Affairs Dir.	Randy Lawson	pddir@kasparradio.com
Sports Director	Shan Sheridan	shan@ccinchamber.org
Accepts PSAs?	yes (contact LeRoy Green, communitybulletinboard@yahoo.com)	

WILO

Web Site	www.wilo.us
Wattage	250
Format	Adult Standards/News/Talk
On-air Hours	5:00 a.m. - 11:00 p.m.

WSHW

Web Site	www.shine99.com
Wattage	50,000
Format	Hot Adult Contemporary
On-air Hours	24/7
Notes	Repeats on 107.5 FM (Zionsville).

FRANKLIN

Atterbury Crier — Daily Journal — Edinburgh Courier

Mailing Address	P.O. Box 699, Franklin IN 46131
Street Address	2575 N. Morton St., Franklin IN 46131
Telephone	317-736-7101
Telephone (toll-free)	888-736-7101
Telephone (news)	317-736-2726
Fax	317-736-2754
Fax (news)	317-736-2766
Fax (advertising)	317-736-2713
Web Site	www.dailyjournal.net
Publishing Company	Home News Enterprises
Publisher	Chuck Wells cwells@dailyjournal.net
Advertising Director	Christina Cosner ccosner@dailyjournal.net

Atterbury Crier

E-mail (news)	amay@dailyjournal.net
Publication Date	Monthly
Circulation	2,700 (free)
Editor	Amy May amay@dailyjournal.net
Managing Editor	Paul Hoffman phoffman@dailyjournal.net
Notes	All editorial content is approved by the U.S. Army. Serves Camp Atterbury (Edinburgh) & Muscatatuck Urban Training Center.

Daily Journal

E-mail (news)	newstips@dailyjournal.net
Publication Date	Daily (Monday-Saturday)
Circulation	16,400 (paid)
Editor	Scarlett Syse syse@dailyjournal.net
Managing Editor	Michele Holtkamp mholtkamp@dailyjournal.net
Sports Editor	Rick Morwick rmorwick@dailyjournal.net

Edinburgh Courier

E-mail (news)	courier@dailyjournal.net
Publication Date	Weekly (Thursday)
Circulation	4,500 (free)
Editor	Paul Hoffman phoffman@dailyjournal.net
Notes	Serves Edinburgh.

See **Franklin Challenger** (Greenwood).

See **Franklin Newspaper** (listed under College Campus/Specialty Publications).

See **K-Love** (listed under National Radio Stations).

WFCI (89.5 FM)

Address	101 Branigin Blvd., Franklin IN 46131
Telephone	317-738-8205
Wattage	1,150
Format	Simulcast of WFYI FM (Indianapolis) during non-academic year and 5:00 a.m. to 7:00 p.m. during academic year. Oldies from 7:00 p.m. to 5:00 a.m. during academic year.
On-air Hours	24/7
Owner	Franklin College
Faculty Advisor	Joel Cramer jcramer@franklincollege.edu
Accepts PSAs?	yes
Notes	Non-commercial station.

WFDM (95.9 FM) — WXLW (950 AM)

Address	645 Industrial Dr., Franklin IN 46131
Telephone	317-736-4040
Fax	317-736-4781
On-air Hours	24/7
Broadcast Company	Pilgrim Communications LLC
General Manager	Randy Tipmore rtipmore@XL950.com
Program Director	Jeremy Beutel jeremy@XL950.com
Sales Manager	Jeremy Bialek jeremy@freedom959.com
Promotions Director	Derek Schultz derek@XL950.com
Accepts PSAs?	yes (contact Jakim Bialek, production@freedom959.com)
Notes	Stations not broadcast local news.

WFDM

Web Site	www.freedom959.com
Wattage	3,300
Format	Conservative Talk
Networks	Metro, Talk Radio Network, ABC, USA

WXLW

Web Site	www.XL950.com
Wattage	5,000
Format	Sports/Talk
Network	Sporting News Radio
Sports Director	Derek Schultz derek@XL950.com

FRENCH LICK

Orange County

Springs Valley Herald

Mailing Address	P.O. Box 311, French Lick IN 47432
Street Address	8481 College St., French Lick IN 47432
Telephone	812-936-9630
Fax	812-936-9559
E-mail	svh@bluemarble.net
Web Site	www.springsvalleyherald.com
Publication Date	Weekly (Wednesday)
Circulation	3,000 (paid)
Publishing Company	Orange County Publishing
Publisher/Editor	Art Hampton
Sports Editor	Dennis Ellis

See **WFIU** (Bloomington).

WFLQ (100.1 FM)

Mailing Address	P.O. Box 100, French Lick IN 47432
Street Address	2593 N. County Road 810 W., West Baden IN 47469
Telephone	812-936-9100
Fax	812-936-9495
E-mail	wflqfm@smithville.net
Web Site	www.wflq.com
Wattage	6,000
Format	Country
Network Affiliations	ABC Country Coast to Coast
On-air Hours	24/7
Broadcast Company	Willtronics Broadcasting
General Mgr./Sales Mgr.	Bill Willis
News Director	Joe Randolph
Program Director	Randall Hamm
Sports Director	Mike Hamilton
Accepts PSAs?	yes (contact Randall Hamm)

GARRETT

DeKalb County

See **Garrett Clipper** (Auburn).

GARY

Lake County

The 411 Newspaper

Address	1130 Camellia Dr., Munster IN 46321
Telephone	219-922-8846
Fax	219-922-9091
E-mail	gary411news@aol.com
Web Site	www.gary411news.com
Publication Date	Weekly (Friday)
Circulation	5,000 (paid/mailed & newsstand)
Owner	Jackie Harris
Notes	Serves Gary.

Gary Crusader Newspaper

Address	1549 Broadway, Gary IN 46407
Telephone	219-885-4357
Fax	219-883-3317
E-mail	garycrusadernews@aol.com
Web Site	www.garycrusadernews.com
Publication Date	Weekly (Thursday)
Circulation	40,000 (paid)
Publisher/Editor	Dorothy Leavell
Managing Editor	David B. Denson
Advertising Manager	John L. Smith
Sports Editor	Lionel Chambers
Notes	Also publishes Chicago Crusader with office in Chicago. Serves African-American community.

WGVE (88.7 FM)

Address	1800 E. 35th Ave., Gary IN 46409
Telephone	219-962-9483
Fax	219-962-3726
E-mail	saritaastevens@yahoo.com
Wattage	2,100
Format	Talk/Music
On-air Hours	24/7
Owner	Gary Community School Corp.
General Manager	Sarita Stevens
Accepts PSAs?	yes
Notes	Non-commercial station.

See **WHNW TV** (South Bend).

GAS CITY

Grant County

Courier — The Giant — Journal-Reporter — Oak Hill Times

Address	787 E. Main St., Gas City IN 46933
Telephone	765-674-0070
Fax	765-674-3496
E-mail	editor1@indy.rr.com
Publication Date	Weekly (Wednesday)
Circulation	15,000-paid (combination of all 4 newspapers)
Publishing Company	The Indiana Newspaper Group
Publisher	Greg LeNeave greg007@comcast.net
Editor	Rachel Kennedy editor1@indy.rr.com
Notes	Courier serves eastern Grant County. Journal-Reporter serves central Grant County. The Giant serves Marion. Oak Hill Times serves western Grant County.

See **Riverside Palaver** (Fairmount).

GOSHEN

Elkhart County

El Puente

Mailing Address	P.O. Box 553, Goshen IN 46527
Street Address	1906 W. Clinton St., Goshen IN 46526
Telephone	574-533-9082
Fax	574-537-0552
E-mail	mail@webelpuente.com
Web Site	www.webelpuente.com
Publication Date	Semi-monthly
Circulation	9,000 (free/newsstand)
Publisher/Executive Editor	Zulma Prieto zulma@webelpuente.com
General Manager	Yizzar Prieto design@webelpuente.com
Advertising Manager	Jimmer Prieto correo@webelpuente.com
Notes	Serves Hispanic community in 14 counties in Michiana area.

See **Goshen College Record** (listed under College Campus/Specialty Publications).

Goshen News

Mailing Address	P.O. Box 569, Goshen IN 46527
Street Address	114 S. Main St., Goshen IN 46526
Telephone	574-533-2151
Telephone (toll-free)	800-487-2151
Fax	574-533-0839
Fax (news)	574-534-8830
E-mail (news)	news@goshennews.com
Web Site	www.goshennews.com
Publication Date	Daily (Sunday-Saturday)
Circulation	17,000 (paid)
Publishing Company	CNHI
Publisher	Jim Kroemer — jim.kroemer@goshennews.com
Executive Editor	Mike Wanbaugh — michael.wanbaugh@goshennews.com
Advertising Manager	Mary Kay Beer — mary.beer@goshennews.com
Sports Editor	Stu Swartz — stu.swartz@goshennews.com

'the Paper'—Bureau

Address	134 S. Main St., Goshen IN 46526
Telephone	574-534-2591
Fax	574-533-4280
E-mail	goshen@the-papers.com
Office Manager	Marilyn Yoder
Notes	Main office in Milford.

WGCS (91.1 FM)

Address	1700 S. Main St., Goshen IN 46526
Telephone	574-535-7488
Fax	574-535-7293
E-mail	globe@goshen.edu
Web Site	www.globeradio.org
Wattage	6,000
Format	AAA/Americana
Network Affiliations	PRI, BBC
On-air Hours	24/7
Broadcast Company	Goshen College Broadcasting Corp.
General Manager	Jason Samuel — jasonks@goshen.edu
Accepts PSAs?	yes (contact Jason Samuel)
Notes	Non-commercial station. Does not broadcast local news.

WKAM (1460 AM)

Address	930 E. Lincoln Ave., Goshen IN 46528
Telephone	574-533-1460
Fax	574-534-3698
Wattage	2,500
Format	Hispanic
On-air Hours	24/7
Owner/General Manager	Ignacio Zepeda
Traffic Manager	Candy Gibbs — candy.lamejor@hotmail.com
Accepts PSAs?	yes (contact Ignacio Zepeda)

WLDC (105.9 FM)

Address	1213 E. Lincoln Ave., Goshen IN 46528
Telephone	574-537-8171
E-mail	avivamiento3a5@yahoo.com
Web Site	www.ministeriossinai.org
Wattage	94
Format	Hispanic
On-air Hours	24/7
Owner	Iglesia Sinai
General Manager	Noe Campos
Accepts PSAs?	yes

GRABILL

Allen County

East Allen Courier

Mailing Address	P.O. Box 77, Grabill IN 46741
Street Address	13720 N. Main St., Grabill IN 46741
Telephone	260-627-2728
Fax	260-627-2519
E-mail	courier@tk7.net
Publication Date	Weekly (Tuesday)
Circulation	7,500 (free & paid)
Publisher/Editor	Waldo Dick

GRANGER

Saint Joseph County

Granger Gazette

Mailing Address	P.O. Box 16, Granger IN 46530
Street Address	14190 Taddington Dr., Granger IN 46530
Telephone	574-277-2679
Fax	574-243-5891
E-mail	grgazette@comcast.net
Publication Date	Monthly
Circulation	11,600 (free/mailed)
Publisher/Editor	Kerry Byler

GREENCASTLE

Banner Graphic

Putnam County

Mailing Address	P.O. Box 509, Greencastle IN 46135
Street Address	100 N. Jackson St., Greencastle IN 46135
Telephone	765-653-5151
Telephone (toll-free)	888-778-8877
Fax	765-653-2063
E-mail (news)	news@bannergraphic.com
Web Site	www.bannergraphic.com
Publication Date	Daily (Monday & Wednesday-Saturday)
Circulation	10,600-paid & free (Monday) & 5,800-paid (Wednesday-Saturday)
Publishing Company	Rust Communications
Publisher	Randy List	rlist2@hotmail.com
General Manager	Daryl Taylor	dtaylor@bannergraphic.com
Editor	Jamie Barrand	jbarrand@bannergraphic.com
Advertising Manager	John York	john@bannergraphic.com
Sports Editor	Caine Gardner	sports@bannergraphic.com

See **The DePauw** (listed under College Campus/Specialty Publications).

WGRE (91.5 FM)

Address	609 S. Locust St., Greencastle IN 46135
Telephone	765-658-4643
Telephone (news)	765-658-4639
Fax	765-658-4693
E-mail	wgre@depauw.edu
E-mail (news)	wgrenews@depauw.edu
Web Site	www.wgre.org
Wattage	800
Format	Alternative
On-air Hours	24/7 (August-May)
Owner	DePauw University
Faculty Advisor	Jeff McCall	jeffmccall@depauw.edu
Accepts PSAs?	yes (contact Program Director)
Notes	Non-commercial, educational station.

WREB—Studio

Address	2468 E. County Rd. 25 N., Greencastle IN 46135
Telephone	765-653-9717
Fax	765-653-6677
E-mail	wreb@originalcompany.com
Notes	Main office in Vincennes.

GREENFIELD

Daily Reporter — Fortville-McCordsville Reporter
New Palestine Reporter

Hancock County

Mailing Address	P.O. Box 279, Greenfield IN 46140
Street Address	22 W. New Rd., Greenfield IN 46140
Telephone	317-462-5528
Telephone (toll-free)	800-528-3717
Telephone (news)	317-467-6022
Fax	317-467-6017
Fax (advertising)	317-467-6009
Web Site	www.greenfieldreporter.com
Publishing Company	Home News Enterprises
Publisher	Randall Shields	rdshields@greenfieldreporter.com
Advertising Manager	John Senger	jsenger@greenfieldreporter.com
TMC/Shopper	Advertiser (weekly)

Daily Reporter

E-mail (news)	editorial@greenfieldreporter.com	
Publication Date	Daily (Tuesday-Saturday)	
Circulation	10,000 (paid)	
Editor	David Hill	dhill@greenfieldreporter.com
Sports Editor	Brian Harmon	bharmon@greenfieldreporter.com

Fortville-McCordsville Reporter

E-mail (news)	news@greenfieldreporter.com	
Publication Date	Weekly (Thursday)	
Circulation	7,000 (free/mailed)	
Editor	Scott Slade	sslade@greenfieldreporter.com
Notes	Serves Fortville, McCordsville & Cumberland.	

New Palestine Reporter

E-mail (news)	news@greenfieldreporter.com	
Publication Date	Weekly (Friday)	
Circulation	6,000 (free/delivered)	
Editor	Scott Slade	sslade@greenfieldreporter.com
Notes	Serves Cumberland & New Palestine.	

Special Edition

Telephone/Fax	317-462-0151
E-mail	specialeditionmonthly@comcast.net
Web Site	www.specialeditionmonthly.com
Publication Date	Monthly
Circulation	9,000
Publishing Company	Special Edition, LLC
Publisher	Kimberly Creech
Notes	Serves Greenfield.

WENS (90.1 FM) — **WJCF** (88.1 FM) — **WRFM** (990 AM) — **WRFM** (89.1 FM)

Mailing Address	P.O. Box 846, Greenfield IN 46140
Street Address	15 Wood St., Greenfield IN 46140
Telephone	317-467-1064
Telephone (toll-free)	877-888-5773
Fax	317-467-1065
E-mail	wjcfradio@aol.com
Web Site	www.wjcfradio.org
Wattage	3,700 (WENS)
	2,700 (WJCF)
	250 (WRFM AM)
	150 (WRFM FM)
Format	True Talk for Indiana (WENS & WRFM AM)
	Contemporary Christian (WJCF)
	True Oldies (WRFM FM)
Network Affiliations	Salem Radio, Moody, Focus on the Family
On-air Hours	24/7
Broadcast Company	Indiana Community Radio Corp.
General Mgr./Sales Mgr.	Jennifer Cox-Hensley lovebeingamommy@aol.com
News/Program Director	Martin Hensley hensleym31@aol.com
Promotions Director	Pat Diemer wjcf_radio@hotmail.com
Accepts PSAs?	yes (contact Martin Hensley)
Notes	Non-commercial stations.
	WENS repeats on 106.9 FM (Fort Branch) and 103.7 FM Rushville).
	WJCF serves Morristown and repeats on 107. 3 FM (Arcadia), 104.5 FM (Connersville), 102.3 FM (Flat Rock), 103.5 FM (Greensburg), 102.9 FM (Mt. Comfort), 101.5 FM (Muncie), 93.9 (New Castle), 98.7 FM &104.9 FM (Rushville), and 96.3 FM (Shelbyville).
	WRFM AM serves Muncie.
	WRFM FM serves Wilkinson and repeats on 105.3 FM (Anderson), 107.1 FM (Connersville), 91.7 FM (Edinburgh), and 104.5 FM (Greensburg).

WRGF (89.7 FM)

Address	810 N. Broadway, Greenfield IN 46140
Telephone	317-477-4603
Fax	317-467-6755
E-mail	wrgf@comcast.net
Wattage	2,000
Format	Eclectic Rock
Network Affiliations	Network Indiana
On-air Hours	24/7
Owner	Greenfield-Central Community Schools
General Manager	Tim Renshaw
Accepts PSAs?	yes (contact Tim Renshaw)
Notes	Non-commercial station.

GREENSBURG

Decatur County

Greensburg Daily News

Address	135 S. Franklin St., Greensburg IN 47240
Telephone	812-663-3111
Telephone (toll-free)	877-253-7758
Fax	812-663-2985
Web Site	www.greensburgdailynews.com
Publication Date	Daily (Monday-Saturday)
Circulation	5,900 (paid)
Publishing Company	CNHI
Publisher	Laura Welborn	laura.welborn@indianamediagroup.com
Managing Editor	Adam Huening	adam.huening@greensburgdailynews.com
Advertising Manager	Keith Wells	keith.wells@indianamediagroup.com
Sports Editor	Nick Gonnella	nick.gonnella@greensburgdailynews.com

---

See **WAUZ** (Columbus).

See **WFIU** (Bloomington).

See **WJCF** and **WRFM FM** (Greenfield).

WTRE (1330 AM)

Mailing Address	P.O. Box 487, Greensburg IN 47240
Street Address	1217 W. Park Rd., Greensburg IN 47240
Telephone	812-663-3000
Fax	812-663-8355
E-mail	wtre@1330wtre.com
Web Site	www.1330wtre.com
Wattage	500
Format	Country
Network Affiliations	Waitt Radio Network
On-air Hours	24/7
Broadcast Company	Reising Radio Partners
News Director	Jennifer McNealy	jmcnealy@1330wtre.com
Sales Manager	Dawn Daugherty-Andrews	ddaugherty@qmix.com
Station Manager	Sandy Biddinger	sbiddinger@1330wtre.com
Accepts PSAs?	yes (contact Sandy Biddinger)

GREENWOOD

Franklin Challenger — Greenwood & Southside Challenger
The Whiteland Times

Street Address	400 E. Main St., Greenwood IN 46142
Telephone	317-888-3376
Fax	317-888-3377
E-mail	news@indychallenger.com
Web Site	www.challengernewspapers.com
Publishing Company	Greenwood Newspapers
Publisher/Editor	Doug Chambers doug@indychallenger.com

Franklin Challenger

Mailing Address	P.O. Box 73, Franklin IN 46131
Publication Date	Weekly (Thursday)
Circulation	3,200 (paid/mailed)
Notes	Serves Franklin.

Greenwood & Southside Challenger

Mailing Address	P.O. Box 708, Greenwood IN 46142
Publication Date	Weekly (Wednesday)
Circulation	4,300 (paid/mailed)

The Whiteland Times

Mailing Address	P.O. Box 242, Whiteland IN 46184
Publication Date	Monthly
Circulation	1,700 (paid/mailed)
Notes	Serves Whiteland.

Indianapolis Star—Metro South Bureau

Address	65 Airport Pkwy., STE 130, Greenwood IN 46143
Telephone	317-444-2700
Fax	317-444-8700
E-mail	starsouth@indystar.com
Web Site	www.indystar.com
Suburban Editor	Kevin Morgan kevin.morgan@indystar.com

WCLJ-TV (Channel 42)

Address	2528 U.S. 31 S., Greenwood IN 46143
Telephone	317-535-5542
Telephone (toll-free)	800-535-5542
Fax	317-535-8584
E-mail	wcljpa@tbn.org
Web Site	www.tbn.org
Network Affiliation	TBN
On-air Hours	24/7
Broadcast Company	Trinity Broadcasting Network
General Manager	Mark Crouch mcrouch@tbn.org
Public Affairs Director	Karen Ward wcljpa@tbn.org
Accepts PSAs?	yes (contact Karen Ward)
Notes	Not-for-profit/non-commercial station. Does not broadcast local news. Serves Bloomington & Indianapolis.

HAGERSTOWN

See **Nettle Creek Gazette** (Cambridge City).

Wayne County

HAMILTON

Steuben County

Hamilton News

Mailing Address	P.O. Box 326, Hamilton IN 46742
Street Address	3950 E. Church St., Hamilton IN 46742
Telephone	260-488-3780
Fax	260-488-4326
E-mail	news@thehamiltonnews.com
Web Site	www.hamiltonnewsonline.com
Publication Date	Weekly (Tuesday)
Circulation	900 (paid)
Publishing Company	Steuben Publishing, Inc.
Editor/Advertising Mgr.	Tracy Thornbrugh

HAMMOND

Lake County

See **Chronicle** (listed under College Campus/Specialty Publications).

WJOB (1230 AM)

Address	6405 Olcott Ave., Hammond IN 46320	
Telephone	219-844-1230	
Telephone (news)	219-844-1416	
Fax	219-989-8516	
Web Site	www.heyregion.com	
Wattage	1,000	
Format	News/Talk/Sports/Music	
On-air Hours	24/7	
Broadcast Company	Vasquez Development	
Owner	Alexis Vasquez Dedlow	alexis@heyregion.com
General Manager	James Dedelow	jed@heyregion.com
Program Director	Michael Stewart	stew@heyregion.com
Sales Manager	Debbie Wargo	debbie@heyregion.com
Office Manager	Vera Mileusnic	vera@heyregion.com
Accepts PSAs?	yes (contact Michael Stewart)	

WPWX (92.3 FM) — WSRB (106.3 FM) — WYCA (102.3 FM)

Address	6336 Calumet Ave., Hammond IN 46324
Telephone	219-933-4455
Fax	219-933-0323
On-air Hours	24/7
Broadcast Company	Crawford Broadcasting
General Manager	Taft Harris
Accepts PSAs?	yes (contact Debra Rhodes)
Notes	Does not broadcast local news.

WPWX

E-mail	power92feedback@crawfordbroadcasting.com	
Web Site	www.power92chicago.com	
Wattage	50,000	
Format	Urban Contemporary	
Program Director	Jay Alan	
Sales Manager	Ruben Cornejo	cornejor@wpwxsales.com

WSRB

E-mail	soulfeedback@crawfordbroadcasting.com	
Web Site	www.soul1063radio.com	
Wattage	4,100	
Format	Urban	
Program Director	Tracie Reynolds	
Sales Manager	Ruben Cornejo	cornejor@wpwxsales.com

WYCA

Web Site	www.wyca1023.com	
Wattage	4,500	
Format	Gospel	
Program Director	Debra Rhodes	
Sales Manager	Darryll King	darryllking@crawfordbroadcasting.com

HARDINSBURG
Washington County

See **WKLO** (Paoli).

HARTFORD CITY
Blackford County

News Times

Mailing Address	P.O. Box 690, Hartford City IN 47348
Street Address	123 S. Jefferson St., Hartford City IN 47348
Telephone	765-348-0110
Telephone (toll-free)	800-435-1842
Fax	765-348-0112
E-mail	ntoffice@comcast.net
E-mail (news)	newstimes@comcast.net
Web Site	www.hartfordcitynewstimes.com
Publication Date	Daily (Monday & Wednesday-Saturday)
Circulation	1,400 (paid)
Publishing Company	Community Media Group
Publisher/Editor	Cynthia Payne
Managing Editor	Robyn Rogers ntrobyn@comcast.net
TMC/Shopper	Red Ball Express (weekly)

HEBRON

Porter County

The Hebron Advertiser

Mailing Address	P.O. Box 2, Hebron IN 46341
Street Address	131 N. Main St., Hebron IN 46341
Telephone	219-996-3142
Fax	219-996-3144
E-mail	sales@russprintshop.com
Web Site	www.hebronadvertiser.com
Publication Date	Weekly (Saturday)
Circulation	21,000 (free/mailed)
Publishing Company	Russ' Print Shop
Publisher	Russ Franzman
Editor	Sue Franzman
Advertising Manager	Chad Franzman

HELMSBURG

Brown County

Our Brown County

Mailing Address	P.O. Box 157, Helmsburg IN 47435
Telephone	812-988-8807
E-mail	ourbrown@bluemarble.net
Web Site	www.ourbrowncounty.com
Publication Date	Bi-monthly
Circulation	20,000 (free/newsstand)
Publishing Company	Singing Pines Projects
Publisher/Editor	Cindy Steele
Notes	Also publishes **INTO ART** (art publication serving Bloomington, Columbus & Nashville) three times per year.

HOBART

Lake County

See **The Chronicle** (Valparaiso).

HOPE

Bartholomew County

Hope Star-Journal

Mailing Address	P.O. Box 65, Hope IN 47246	
Street Address	308 Jackson St., Hope IN 47246	
Telephone	812-546-4940	
Fax	812-546-4944	
E-mail	news@hopestarjournal.com	
Web Site	www.hopestarjournal.com	
Publication Date	Weekly (Thursday)	
Circulation	1,000 (paid)	
Publishing Company	Indiana News Media LLC	
Publisher/Executive Editor	Larry Simpson	lsimpson@hopestarjournal.com
Advertising Manager	Stephanie Shoaf	ads@hopestarjournal.com

HOWE

LaGrange County

WHWE (89.7 FM)

Mailing Address	P.O. Box 240, Howe IN 46746
Street Address	5755 N. State Rd. 9, Howe IN 46746
Telephone	260-562-2131
Fax	260-562-3678
E-mail	sclark@howemilitary.com
E-mail (news)	kcalvert@howemilitary.com
Web Site	www.howemilitary.com
Wattage	100
Format	Educational
On-air Hours	8:00 a.m.-4:00 p.m. (Monday-Friday)
Owner	Howe Military School
General Manager	Steve Clark sclark@howemilitary.com
Program Director	Kristen Calvert kcalvert@howemilitary.com
Accepts PSAs?	yes (contact Kristen Calvert)
Notes	Non-commercial station.

See **WQKO** (Valparaiso).

HUNTERTOWN

Allen County

Northwest News

Mailing Address	P.O. Box 663, Huntertown IN 46748
Street Address	15605 Lima Rd., Huntertown IN 46748
Telephone	260-637-9003
Fax	260-637-8598
E-mail	nweditor@app-printing.com
Web Site	www.app-printing.com
Publication Date	Weekly (Wednesday)
Circulation	1,400 (paid)
Publishing Company	All Printing & Publications
Publisher	Robert Allman
Editor	Ryan Schwab

HUNTINGBURG

Dubois County

Huntingburg Press

Mailing Address	P.O. Box 260, Huntingburg IN 47542
Street Address	600 E. 6th St., STE B, Huntingburg IN 47542
Telephone	812-683-5899
Fax	812-683-5897
E-mail	hbgpress@insightbb.com
Publication Date	Weekly (Friday)
Publishing Company	Kentucky Publishing
Publisher	Greg LeNeave greg007@comcast.net
Editor	Bruce Simmons

WAXL (103.3 FM) — WBDC (100.9 FM)

Mailing Address	P.O. Box 330, Huntingburg IN 47542	
Street Address	501 Old State Rd. 231, Huntingburg IN 47542	
Telephone (toll-free)	800-522-1033	
Fax	812-683-5891	
On-air Hours	24/7	
Broadcast Company	DC Broadcasting	
General Manager	Bill Potter	gm@dcbroadcasting.com
News Director	Chris Hill	news@wbdc.us
Program/Public Affairs	Joe Lacay	onair@wbdc.us
Sports Director	Kurt Gutgsell	
Sales Manager	Ron Spaulding	sales@dcbroadcasting.com
Promotions Director	Paul Knies	
Accepts PSAs?	yes (contact Joe Lacay)	

WAXL

Telephone	812-683-1215
E-mail	mailbox@waxl.us
E-mail (news)	news@waxl.us
Web Site	www.waxl.us
Wattage	3,000
Format	Adult Contemporary
Networks	ABC, Network Indiana, Brownfield Ag
Notes	Serves Santa Claus.

WBDC

Telephone	812-683-4144
E-mail	mailbox@wbdc.us
E-mail (news)	news@wbdc.us
Web Site	www.wbdc.us
Wattage	25,000
Format	Country
Networks	CNN News, Brownfield Ag, Dial Global, AP
Notes	Serves Huntingburg & Jasper.

HUNTINGTON

Huntington County

Herald-Press

Address	7 N. Jefferson St., Huntington IN 46750	
Telephone	260-356-6700	
Fax	260-356-9026	
E-mail	hpnews@h-ponline.com	
Web Site	www.h-ponline.com	
Publication Date	Daily (Sunday-Friday)	
Circulation	5,200 (paid)	
Publishing Company	Paxton Media Group	
Publisher	Andy Eads	aeads@h-ponline.com
Editor	Rebecca Sandlin	rsandlin@h-ponline.com
Advertising Manager	June Whittamore	jwhittamore@h-ponline.com
TMC/Shopper	Current (weekly)	

Huntington County TAB

Mailing Address	P.O. Box 391, Huntington IN 46750
Street Address	1670 Etna Ave., Huntington IN 46750
Telephone	260-356-1107
Fax	260-356-1177
E-mail (news)	tabnewsroom@comcast.net
E-mail (display advertising)	tabads@comcast.net
E-mail (classified advertising)	tabclass@comcast.net
Web Site	www.huntingtoncountytab.com
Publication Date	Semi-weekly (Monday & Thursday)
Circulation	15,000 (free/delivered)
Publishing Company	Huntington TAB Inc.
Publisher/Advertising Mgr.	Russ Grindle
Publisher/Managing Editor	Scott Trauner
News Editor	Cindy Klepper

See **WBZQ** (Fort Wayne).

WQHU (105.5 FM)

Address	2303 College Ave., Huntington IN 46750
Telephone	260-359-4281
Fax	260-359-4249
Web Site	www.wqhu.net
Wattage	100
Format	Rock Alternative/Independent
Owner	Huntington University
Faculty Advisor	Lance Clark lclark@huntington.edu
Accepts PSAs?	yes
Notes	Non-commercial station.

WVSH (91.9 FM)

Address	450 MacGahan St., Huntington IN 46750
Telephone	260-356-2019
Wattage	920
Format	Contemporary Hit Radio/Variety
Owner	Huntington North High School
Faculty Advisor	Nick Altman naltman@hccsc.k12.in.us
Accepts PSAs?	yes
Notes	Non-commercial station.

INDIANAPOLIS

Marion County

Associated Press

Address	251 N. Illinois St., STE 1600, Indianapolis IN 46204
Telephone	317-639-5501
Telephone (State House)	317-631-8629
Fax	317-638-4611
E-mail	indy@ap.org
Web Site	www.ap.org
Bureau Chief	Keith Robinson
State House Correspondent	Mike Smith

Hoosier Ag Today

Address	P.O. Box 34236, Indianapolis IN 46234	
Telephone	317-247-9360	
Fax	317-247-9380	
President	Gary Truitt	gtruitt@hoosieragtoday.com
V. P. Operations	Andy Eubank	aeubank@hoosieragtoday.com
Chief Financial Officer	Kathleen Truitt	ktruitt@hoosieragtoday.com
Traffic Manager	Beth Carper	traffic@hoosieragtoday.com
Notes	Agriculture network serving Indiana radio stations.	

Inside Indiana Business with Gerry Dick

Address	1630 N. Meridian St., STE 400, Indianapolis IN 46202
Telephone	317-275-2010
E-mail	newsletter@growindiana.net
Web Site	www.insideindianabusiness.com
Notes	News service covering Indiana business.

Network Indiana

Address	40 Monument Cir., STE 400, Indianapolis IN 46204	
Telephone	317-637-4638	
Fax	317-684-2008	
Web Site	www.networkindiana.com	
General Manager	Charlie Morgan	charliemorgan@indy.emmis.com
Operations Manager	John Emerson	jemerson@emmis.com
General Sales Manager	Eric Wunnenberg	ewunnenberg@wibc.emmis.com
Public Affairs Director	Matt Hibbeln	mhibbeln@wibc.emmis.com
Notes	Network serving Indiana radio stations.	

atCarmel Community Newsletter — atGeist Community Newsletter
Mailing Address	P.O. Box 36097, Indianapolis IN 46236
Telephone	317-823-5060
Fax	317-536-3030
Publication Date	monthly
Publishing Company	Britt Interactive LLC

atCarmel Community Newsletter
Web Site	www.atCarmel.com	
Circulation	20,000 (free/mailed & newsstand)	
Publisher	Tom Britt	tom@atCarmel.com
Editor	Laura Gates	laura@atCarmel.com
Notes	Serves Carmel.	

atGeist Community Newsletter
Web Site	www.atGeist.com	
Circulation	13,500 (free/mailed & newsstand)	
Publisher	Tom Britt	tom@atGeist.com
Editor	Laura Gates	laura@atGeist.com
Notes	Serves Indianapolis & Fishers.	

See **Butler Collegian** (listed under College Campus/Specialty Publications).

See **Criterion** (listed under Religion/Specialty Publications).

Evansville Courier & Press—Bureau
Address	200 W. Washington St., RM M-7, Indianapolis IN 46204	
Telephone	317-631-7405	
Fax	317-631-0144	
E-mail	bradnere@courierpress.com	
Web Site	www.courierpress.com	
Bureau Chief	Eric Bradner	bradnere@courierpress.com
Notes	Main office in Evansville.	

Franklin Township Informer
Address	8822 Southeastern Ave., Indianapolis IN 46239
Telephone	317-862-1774
Fax	317-862-1775
E-mail	ftinformer@sbcglobal.net
Web Site	www.ftcivicleague.org
Publication Date	Weekly (Wednesday)
Circulation	2,000 (paid/mailed & newsstand)
Owner	Franklin Township Civic League
Editor	Kasie L. Foster
Managing Editor	Marcus Olibo
Advertising Manager	Andy Wilkins
Notes	Serves Franklin Township.

Indiana Herald

Mailing Address	P.O. Box 88449, Indianapolis IN 46208
Street Address	2170 N. Illinois St., Indianapolis IN 46202
Telephone	317-923-8291
Fax	317-923-8292
E-mail	herald1@earthlink.net
Web Site	www.indianaherald.com
Publication Date	Weekly (Thursday)
Circulation	15,000 (paid)
Publisher/Editor	Mary Tandy
Advertising Manager	Bernice Matheson
Sports Editor	Leroy Bryant
Notes	Serves African-American community.

Indiana Weekender

Mailing Address	P.O. Box 199137, Indianapolis IN 46219
Street Address	6433 E. Washington St., STE 155, Indianapolis IN 46219
Telephone	317-322-3315
Fax	317-322-3321
E-mail	indianaweekender@aol.com
E-mail (news)	iweekender@sbcglobal.net
Web Site	www.indianaweekender.com
Publication Date	Bi-weekly
Circulation	150,000 (free/delivered & newsstand)
Owner/Publisher/Editor	G. Thomas Wills
Managing Editor/Ad Mgr.	Verla Wills

See **Indianapolis Business Journal** (listed under Business/Specialty Publications).

See **Indianapolis Dine** (listed under Dining/Specialty Publications).

Indianapolis Monthly

Address	40 Monument Circle, STE 100, Indianapolis IN 46204	
Telephone	317-237-9288	
Fax (news)	317-683-2010	
Fax (advertising)	317-684-8356	
Web Site	www.indianapolismonthly.com	
Publication Date	Monthly	
Circulation	45,000 (paid)	
Publishing Company	Emmis Communications	
Publisher	Keith Phillips	keith@indymonthly.emmis.com
Editor	David Zivan	dzivan@indymonthly.emmis.com
Executive Editor	Amy Wimmer Schwarb	awschwarb@indymonthly.emmis.com
Managing Editor	Kim Hannel	khannel@indymonthly.emmis.com

Indianapolis Recorder

Mailing Address	P.O. Box 18499, Indianapolis IN 46218
Street Address	2901 N. Tacoma Ave., Indianapolis IN 46218
Telephone	317-924-5143
Fax (news)	317-924-5148
Fax (advertising)	317-921-6653
E-mail (news)	newsroom@indyrecorder.com
Web Site	www.indianapolisrecorder.com
Publication Date	Weekly (Thursday)
Circulation	22,000 (paid)
Owner	William Mays
Publisher	Carolene Mays
Editor	Shannon Williams shannonw@indyrecorder.com
Sports Editor	Jessica Williams-Gibson
Notes	Serves African-American community.

Indianapolis Star

Mailing Address	P.O. Box 145, Indianapolis IN 46206
Street Address	307 N. Pennsylvania St., Indianapolis IN 46204
Telephone	317-444-4000
Telephone (toll-free)	800-669-7827
Telephone (news)	317-444-6000
Fax (news)	317-444-6600
Fax (advertising)	317-444-7300
E-mail (news)	startips@indystar.com
Web Site	www.indystar.com
Publication Date	Daily (Sunday-Saturday)
Circulation	246,800-paid (Monday-Friday); 216,300-paid (Saturday); 321,100-paid (Sunday)
Publishing Company	Gannett Co. Inc.
Publisher	Michael Kane michael.kane@indystar.com
Editor	Dennis Ryerson dennis.ryerson@indystar.com
Sports Editor	Jim Lefko jim.lefko@indystar.com
Opinions Editor	Tim Swarens tim.swarens@indystar.com
Notes	Bureaus in Avon, Fishers & Greenwood.

See **Indianapolis Tennis Magazine** (listed under Sports/Specialty Publications).

See **Indianapolis Woman Magazine** (listed under Women/Specialty Publications).

See **Indy Tails Pet Magazine** (listed under Pets/Specialty Publications).

See **Indy's Child Parenting Magazine** (listed under Parenting/Specialty Publications).

See **IUPUI Student Media** (listed under College Campus/Specialty Publications).

Journal Gazette—Bureau

Address	200 W. Washington St., RM M-8, Indianapolis IN 46204
Telephone	317-686-0901
Fax	317-636-1124
E-mail	nkelly@jg.net
Web Site	www.journalgazette.net
Bureau Chief	Niki Kelly nkelly@jg.net
Notes	Main office in Fort Wayne.

La Ola Latino-Americana

Mailing Address	P.O. Box 22056, Indianapolis IN 46222
Street Address	4670 W. Washington St., Indianapolis IN 46241
Telephone	317-822-0345
Telephone (toll-free)	877-LEA-LOLA
Fax	317-822-0344
E-mail	info@laolalatinaindy.com
E-mail (advertising)	ads@laolalatinaindy.com
Web Site	www.laolalatinaindy.com
Publication Date	Bi-weekly (Friday)
Circulation	12,000 (free)
Publisher/Editor	Ildefonso Carbajal
Notes	Serves Hispanic community.

La Voz de Indiana

Address	2911 W. Washington St., Indianapolis IN 46222
Telephone	317-423-0957
Telephone (toll-free)	888-864-2301
Telephone (news)	317-423-0953
Fax	317-423-0956
Fax (advertising)	317-423-0957
E-mail	voz2@cs.com
E-mail (news)	press@lavozdeindiana.com
Web Site	www.lavozdeIndiana.com
Publication Date	Bi-weekly (Wednesday)
Circulation	20,000 (free/newsstand)
Publisher/Editor/Ad Mgr.	Liliana Hamnik
Managing Editor	Claudia Cuartas
Vice President	Jose Gonzalez
Notes	Bilingual newspaper serving the Hispanic community.

Louisville Courier-Journal—Bureau

Address	200 W. Washington St., RM M11, Indianapolis IN 46204
Telephone	317-444-2780
E-mail	Lstedman@courier-journal.com
Web Site	www.courier-journal.com
Bureau Chief	Lesley Stedman Weidenbener Lstedman@courier-journal.com
Notes	Main office in Louisville, Kentucky (800-765-4011)

Metromix Indianapolis

Mailing Address	P.O. Box 145, Indianapolis IN 46206	
Street Address	307 N. Pennsylvania St., Indianapolis IN 46204	
Telephone	317-444-6267	
Fax	317-444-8999	
E-mail	indianapolis@metromix.com	
E-mail (news)	amanda.kingsbury@indystar.com	
Web Site	www.indy.metromix.com	
Publication Date	Weekly (Thursday)	
Circulation	40,000 (free/newsstand)	
Publishing Company	Gannett Co. Inc.	
Publisher	Michael Kane	michael.kane@indystar.com
General Manager	Joey Montgomery	jmontgomery@metromix.com
Editor	Amanda Kingsbury	amanda.kingsbury@indystar.com
Online Producer	Amy Bartner	abartner@metromix.com
Mktg/Promotions Coord.	Adriane Courtney	acourtney@metromix.com
Notes	Entertainment publication.	

NUVO Newsweekly

Address	3951 N. Meridian St., STE 200, Indianapolis IN 46208	
Telephone	317-254-2400	
Fax	317-254-2405	
Fax (advertising)	317-808-4636	
E-mail	nuvo@nuvo.net	
E-mail (news)	editors@nuvo.net	
Web Site	www.nuvo.net	
Publication Date	Weekly (Wednesday)	
Circulation	40,000 (free/newsstand)	
Owner/Publisher/Editor	Kevin McKinney	kmckinney@nuvo.net
Managing Editor	Jim Poyser	jpoyser@nuvo.net
Advertising Manager	Josh Schuler	jschuler@nuvo.net
News/Sports Editor	Laura McPhee	lmcphee@nuvo.net

See **The Reflector** (listed under College Campus/Specialty Publications).

The Times of Northwest Indiana—Bureau

Address	150 W. Market St., STE 135, Indianapolis IN 46204	
Telephone	317-637-9078	
Fax	317-637-9210	
State House Bureau Chief	Daniel Carden	daniel.carden@nwi.com
Notes	Main office in Munster.	

West Indianapolis Community News — West Side Community News

Address	608 S. Vine St., Indianapolis IN 46241
Telephone	317-241-7363
Fax	317-240-6397
E-mail	commnews@in-motion.net
Publication Date	Weekly (Wednesday)
Circulation	25,000-free (combined)
Publishing Company	Community Papers Inc.
Publisher	Jackie F. Deppe

See **West Side Messenger** (Speedway).

See **Westside Flyer** (Avon).

WBDG (90.9 FM)

Address	1200 N. Girls School Rd., Indianapolis IN 46214
Telephone	317-227-4122
Fax	317-243-5506
E-mail	wbdgindy@yahoo.com
Web Site	www.wayne.k12.in.us/bdwbdg
Wattage	400
Format	Variety
On-air Hours	24/7
Owner	MSD of Wayne Township
General Manager	Jon Easter jon.easter@wayne.k12.in.us
Public Affairs Director	Emerson Allen emerson.allen@wayne.k12.in.us
Accepts PSAs?	yes (contact Jon Easter)
Notes	Non-commercial station. Does not broadcast local news.

WBRI (1500 AM)

Address	4802 E. 62nd St., Indianapolis IN 46220
Telephone	317-255-5484
Fax	317-255-8592
E-mail	wbri@wilkinsradio.com
Wattage	5,000
Format	Christian
On-air Hours	6:00 a.m. - 7:00 p.m.
Broadcast Company	Wilkins Communications
General Manager	Keith Smiley
Accepts PSAs?	yes (church-related only)
Notes	Does not broadcast local news.

WEDJ (107.1 FM) — WSYW (810 AM)

Address	1800 N. Meridian St., STE 603, Indianapolis IN 46202
Telephone	317-924-1071
Fax	317-924-7766
Fax (sales)	317-924-7765
Web Site	www.wedjfm.com
Wattage	6,000 (WEDJ)
	250 (WSYW)
Format	Regional Mexican (WEDJ)
	Spanish Variety (WSYW)
On-air Hours	24/7 (WEDJ)
	sunrise — sunset (WSYW)
Broadcast Company	Continental Broadcast Group LLC
General Mgr./Sales Mgr.	Bart Johnson bart@wedjfm.com
Program Director	Manuel Sepulveda manny@wedjfm.com
Traffic Director	Norma Flores norma@wedjfm.com
Accepts PSAs?	yes (contact Manuel Sepulveda)

WEDM (91.1 FM)

Address	9651 E. 21st St., Indianapolis IN 46229
Telephone	317-532-6301
E-mail	dhenn@warren.k12.in.us
Wattage	180
On-air Hours	24/7
Owner	MSD of Warren Township
General Manager	Dan Henn dhenn@warren.k12.in.us
Accepts PSAs?	yes (contact Dan Henn)
Notes	Non-commercial station.

WFBQ (94.7 FM) — WNDE (1260 AM) — WRZX (103.3 FM)

Address	6161 Fall Creek Rd., Indianapolis IN 46220
Telephone	317-257-7565
Fax	317-253-6501
On-air Hours	24/7
Broadcast Company	Clear Channel Radio
General Manager	Rick Green rickgreen@clearchannel.com
Sales Manager	Tom Doran tomdoran@clearchannel.com
Sports Director	John Michael jmv@wnde.com
Public Affairs Director	Don Stuck donstuck@clearchannel.com
Accepts PSAs?	yes (contact Don Stuck)

WFBQ

Web Site	www.wfbq.com
Wattage	58,000
Format	Classic Rock
News Director	Kristi Lee kristilee@clearchannel.com
Program Director	Buzz Casey buzzcasey@clearchannel.com

WNDE

Web Site	www.wnde.com
Wattage	5,000
Format	Sports
Network	Fox Sports
Program Director	Buzz Casey buzzcasey@clearchannel.com

WRZX

Web Site	www.x103.com
Wattage	38,000
Format	Alternative
News Director	Don Stuck donstuck@clearchannel.com
Program Director	Lenny Diana ld@x103.com

WFMS (95.5 FM) — WJJK (104.5 FM) — WRWM (93.9 FM)

Address	6810 N. Shadeland Ave., Indianapolis IN 46220
Telephone	317-842-9550
Telephone (news)	317-558-7107
Fax	317-921-1996
Fax (sales)	317-577-3361
E-mail (news)	karyn.sullyvan@cumulus.com
On-air Hours	24/7
Broadcast Company	Cumulus Media Partners
General Manager	Chris Wheat chris.wheat@cumulus.com
News/Public Affairs Dir.	Karyn Sullyvan karyn.sullyvan@cumulus.com
Accepts PSAs?	yes—but no prerecorded PSAs (contact Karyn Sullyvan)
Notes	Local news produced by WTHR TV.

WFMS

E-mail	wfms@indyradio.com	
Web Site	www.wfms.com	
Wattage	50,000	
Format	Country	
Program Director	Vicki Murphy	vicki.murphy@cumulus.com
Sales Manager	Chris Wheat	chris.wheat@cumulus.com
Promotions Director	Lisa Juillerat	lisa.juillerat@cumulus.com

WJJK

E-mail	wjjk@indyradio.com	
Web Site	www.1045wjjk.com	
Wattage	50,000	
Format	Classic Hits	
Program Director	Steve Cannon	steve.cannon@cumulus.com
Sales Manager	Michele Kiefer	michele.kiefer@cumulus.com
Promotions Director	Anna Fraser	anna.fraser@cumulus.com

WRWM

E-mail	warm939@indyradio.com	
Web Site	www.i94hits.com	
Wattage	25,000	
Format	Top 40	
Program Director	Jeff Andrews	jeff.andrews@cumulus.com
Sales Manager	Michele Kiefer	michele.kiefer@cumulus.com
Promotions Director	Anna Fraser	anna.fraser@cumulus.com

WFNI (1070 AM) — WLHK (97.1 FM) — WYXB (105.7 FM)

Address	40 Monument Cir., STE 600, Indianapolis IN 46204
On-air Hours	24/7
Broadcast Company	Emmis Communications
General Manager	Charlie Morgan — charliemorgan@indy.emmis.com
Operations Manager	Bob Richards — brichards@indy.emmis.com
Director of Sales	Tracey Bean — tbean@indy.emmis.com
National Sales Manager	Patty England — pengland@indy.emmis.com
Community Outreach Mgr.	Katherine Simons — ksimons@indy.emmis.com

WFNI

Telephone	317-261-1070
Telephone (news)	317-637-6397
Telephone (studio)	317-239-1070
Fax	317-684-2095
Fax (news)	317-684-2017
E-mail	webmaster@1070thefan.com
Web Site	www.1070thefan.com
Wattage	50,000
Format	Sports/ESPN
Networks	ESPN, Network Indiana
Program Director	Kent Sterling — kts@indy.emmis.com
Sales Manager	Eric Wunnenberg — ewunnenberg@wibc.emmis.com
Promotions Director	Susan Wells — swells@indy.emmis.com
Executive Producer	Michael Grady — mgrady@indy.emmis.com
Accepts PSAs?	yes (contact Michael Grady)

WLHK

Telephone	317-266-9700
Fax	317-684-2095
Web Site	www.hankfm.com
Wattage	50,000
Format	Country
Program Director	Bob Richards — brichards@indy.emmis.com
Sales Manager	Taja Graham — tgraham@indy.emmis.com
Public Affairs Director	Sarah Harris — sharris@indy.emmis.com
Promotions Director	Susan Wells — swells@indy.emmis.com
Accepts PSAs?	yes (contact Bob Richards)

WYXB

Telephone	317-684-1057
Fax	317-684-2016
Web Site	www.b1057.com
Wattage	50,000
Format	Adult Contemporary
Program Director	Bob Richards — brichards@indy.emmis.com
Public Affairs Director	Sarah Harris — sharris@indy.emmis.com
Promotions Director	Katie Webber — kwebber@indy.emmis.com
Accepts PSAs?	yes (contact Bob Richards)

WFYI (90.1 FM)

Address	1630 N. Meridian St., Indianapolis IN 46202
Telephone	317-636-2020
Fax	317-283-6645
E-mail (news)	mhartnett@wfyi.org
Web Site	www.wfyi.org
Wattage	10,000
Format	News/Culture
Network Affiliations	NPR, PRI, BBC
On-air Hours	24/7
Broadcast Company	Metropolitan Indianapolis Public Broadcasting
General Manager	Richard Miles — rmiles@wfyi.org
News Director	Mary Hartnett — mhartnett@wfyi.org
News Anchor/Producer	Sharon Alseth — salseth@wfyi.org
Corp. Development Mgr.	Ian Hall — ihall@wfyi.org
Accepts PSAs?	no
Notes	Non-commercial station. Simulcasts on WFCI (Franklin) and WNDY (Crawfordsville).

See **WGNR AM & FM** (Anderson).

WHHH (96.3 FM) — **WNOU** (100.9 FM) — **WTLC** (1310 AM) — **WTLC** (106.7 FM)

Address	21 E. Saint Joseph St., Indianapolis IN 46204
Telephone	317-266-9600
Fax	317-328-3860
Fax (sales)	317-328-3870
On-air Hours	24/7
Broadcast Company	Radio One
General Manager	Chuck Williams cwilliams@radio-one.com
Public Affairs Director	Amos Brown abrown@radio-one.com
Accepts PSAs?	yes (contact Amos Brown)

WHHH

Web Site	www.hot963.com	
Wattage	6,000	
Format	Mainstream Urban	
News Director	Terri Durrett	tdurrett@radio-one.com
Program Director	Brian Wallace	bwallace@radio-one.com
Sales Manager	Nikki Wills	nwills@radio-one.com
Promotions Director	Shannon Joseph	sjoseph@radio-one.com

WNOU

Web Site	www.radionow1009.com	
Wattage	6,000	
Format	Contemporary Hit Radio/Top 40	
News Director	Shy Holder	sholder@radio-one.com
Program Director	RAYNE	rayne@radio-one.com
Sales Manager	Nikki Wills	nwills@radio-one.com
Promotions Director	Annie Bogigian	abogigian@radio-one.com

WTLC AM

Web Site	www.1310thelight.com	
Wattage	5,000 (day); 1,000 (night)	
Format	Praise/Community Talk	
News Director	Amos Brown	abrown@radio-one.com
Program Director	Kris Rhaye	khenderson@radio-one.com
Sales Manager	Brian Harrington	bharrington@radio-one.com
Promotions Director	Shannon Joseph	sjoseph@radio-one.com

WTLC FM

Web Site	www.wtlc.com	
Wattage	6,000	
Format	Urban Contemporary	
News Director	Terri Durrett	tdurrett@radio-one.com
Program Director	Brian Wallace	bwallace@radio-one.com
Sales Manager	Brian Harrington	bharrington@radio-one.com
Promotions Director	Shannon Joseph	sjoseph@radio-one.com

WIBC (93.1 FM)

Address	40 Monument Cir., STE 400, Indianapolis IN 46204
Telephone	317-266-9422
Telephone (news)	317-637-6397
Telephone (studio)	317-239-9393
Fax	317-684-2095
Fax (news)	317-684-2017
E-mail (news)	news@wibc.com
Web Site	www.wibc.com
Wattage	50,000
Format	News/Talk
Network Affiliations	Network Indiana
On-air Hours	24/7
Broadcast Company	Emmis Communications
General Manager	Charlie Morgan charliemorgan@indy.emmis.com
Program Director	Kent Sterling kts@indy.emmis.com
Sales Manager	Eric Wunnenberg ewunnenberg@wibc.emmis.com
Executive Producer	Matt Hibbeln mhibbeln@wibc.emmis.com
Accepts PSAs?	yes (contact Charlie Dee, charlied@wibc.emmis.com)

WICR (88.7 FM)

Address	1400 E. Hanna Ave., Indianapolis IN 46227
Telephone	317-788-3280
Fax	317-788-3490
E-mail	wicr@uindy.edu
Web Site	www.wicr.uindy.edu
Wattage	5,000
Format	Classical/Jazz
Network Affiliations	PRI
On-air Hours	24/7
Owner	University of Indianapolis
General Manager	Scott Uecker suecker@uindy.edu
Accepts PSAs?	yes (contact wicr@uindy.edu)
Notes	Non-commercial station.

See **WIKL/K-Love** (listed under National Radio Stations).

WITT (91.9 FM)

Mailing Address	P.O. Box 20563, Indianapolis IN 46220
Street Address	6218 Kingsley Dr., Indianapolis IN 46222
Telephone (toll-free)	877-401-3851
E-mail	radio@919witt.org
Web Site	www.919witt.org
Wattage	6,000
Format	Variety
On-air Hours	24/7
General Manager	Jim Walsh jwalsh@919witt.org
Accepts PSAs?	yes (contact walshvid@netdirect.net)
Notes	Non-commercial station. Does not broadcast local news.

WJEL (89.3 FM)

Address	1901 E. 86th St., Indianapolis IN 46240
Telephone	317-259-5278
Fax	317-259-5298
E-mail	rhendrix@msdwt.k12.in.us
Web Site	www.wjel.com
Wattage	1,000
Format	Contemporary Hit Radio
Network Affiliations	ABC News
On-air Hours	24/7
Owner	MSD of Washington Township
General Manager	Rob Hendrix rhendrix@msdwt.k12.in.us
Public Affairs/Promotions	Tyler Hindman thindman@msdwt.k12.in.us
Accepts PSAs?	yes (contact Rob Hendrix)
Notes	Non-commercial station.

See **WKLU/K-Love** (listed under National Radio Stations).

WNTR (107.9 FM) — WXNT (1430 AM) — WZPL (99.5 FM)

Address	9245 N. Meridian St., STE 300, Indianapolis IN 46260
Telephone	317-816-4000
Telephone (news)	317-816-4037
Fax	317-816-4030
Fax (news)	317-816-4060
On-air Hours	24/7
Broadcast Company	Entercom Indianapolis
General Manager	Jennifer Skjodt jskjodt@entercom.com
Operations Director	Scott Sands ssands@entercom.com
Sales Manager	Erika Estridge eestridge@entercom.com
Public Affairs Director	Kelli Jack kjack@entercom.com
Marketing Director	Toni Moore tmoore@entercom.com
Accepts PSAs?	yes (contact Scott Manning via fax)

WNTR

Web Site	www.1079thetrack.com
Wattage	50,000
Format	"The Track"/Adult Contemporary

WXNT

E-mail (news)	wxntnews@entercom.com
Web Site	www.newstalk1430.com
Wattage	5,000
Format	News/Talk
Network	Fox

WZPL

Web Site	www.wzpl.com
Wattage	50,000
Format	Hot Adult Contemporary

WNTS (1590 AM)

Address	3745 W. Washington St., Indianapolis IN 46241
Telephone	317-472-7137
Fax	317-472-7138
Wattage	5,000
Format	Regional Mexican
On-air Hours	24/7
Broadcast Company	Davidson Media Group
General Manager	Ruben Pazmino	rpazmino@aol.com
News Director	Lupita Morales	lupitalaley1590@yahoo.com
Program Director	Mayraelisa Arroyo	mayraelisa00@hotmail.com
Sports Director	Alfonso Romero
Sales Manager	Mark Clarke	lapoderosa1590@aol.com
Accepts PSAs?	yes (contact Mayraelisa Arroyo)

WRFT (91.5 FM)

Address	6215 S. Franklin Rd., Indianapolis IN 46259
Telephone	317-803-5552
Fax	317-862-7262
E-mail	steve.george@ftcsc.k12.in.us
Web Site	www.ftcsc.k12.in.us
Wattage	130
Format	80's and Top 40
On-air Hours	24/7
Owner	Franklin Township Community School Corp.
General Manager	Steve George	steve.george@ftcsc.k12.in.us
Accepts PSAs?	yes
Notes	Non-commercial station.

WSPM (89.1 FM)

Address	3500 DePauw Blvd., STE 2085, Indianapolis IN 46268
Telephone	317-870-8400
Telephone (toll-free)	866-883-8400
Fax	317-870-8404
Web Site	www.catholicradioindy.org
Wattage	22,000
Format	Catholic
Network Affiliations	EWTN
On-air Hours	24/7
Broadcast Company	Hoosier Broadcasting Corp.
General Manager	Jim Ganley	jim@catholicradioindy.org
Accepts PSAs?	yes
Notes	Non-commercial station.

WTTS—Sales Office

Address	407 Fulton St., Indianapolis IN 46202
Telephone	317-972-9887
Fax	317-972-9886
Sales Manager	Daryl McIntire	mac@wttsfm.com
Notes	Main office in Bloomington.

See **WCLJ-TV** (Greenwood).

WDNI TV (Channel 19)

Address	21 E. Saint Joseph St., Indianapolis IN 46204
Telephone	317-266-9600
Fax	317-328-3870
Web Site	www.imc.tv
Network Affiliation	Independent
On-air Hours	24/7
Broadcast Company	Radio One
General Manager	Chuck Williams cwilliams@radio-one.com
General Sales Manager	Ian Banks ibanks@radio-one.com
Program Director	Dan McNeal dmcneal@radio-one.com
Promotions Director	Annie Bogigian abogigian@radio-one.com
Accepts PSAs?	yes (contact Dan McNeal)
Notes	Does not broadcast local news.

WFYI TV (Channels 20.1/WFYI 1; 20.2/WFYI 2/V-me; & 20.3/WFYI 3)

Address	1630 N. Meridian St., Indianapolis IN 46202
Telephone	317-636-2020
Fax	317-283-6645
Web Site	www.wfyi.org
Network Affiliation	PBS
On-air Hours	24/7
Broadcast Company	Metropolitan Indianapolis Public Broadcasting
General Manager	Lloyd Wright lwright@wfyi.org
Program Director (WFYI 1)	Aundrea Hart ahart@wfyi.org
Program Dir. (WFYI 2 & 3)	Alan Carmack acarmack@wfyi.org
Corp. Development Mgr.	Ian Hall ihall@wfyi.org
Media Relations Director	Lori Plummer lplummer@wfyi.org
Accepts PSAs?	no
Notes	Non-commercial station. Does not broadcast local news.

WIPX TV (Channels 63.1/main; 63.2/qubo; 63.3/life; & 63.4/religion)

Address	2441 Production Dr., STE 104, Indianapolis IN 46241
Telephone	317-486-0633
Fax	317-486-0298
Web Site	www.ionline.tv
Network Affiliation	Ion Media Network
On-air Hours	24/7
Broadcast Company	Ion Media Network
Station Operations Mgr.	John Kowalke johnkowalke@ionmedia.tv
Local Sales Manager	Robert Getze robertgetze@ionmedia.tv
Traffic Manager	Terri Durrett terridurrett@ionmedia.tv
Accepts PSAs?	yes (contact Terri Durrett)
Notes	Does not broadcast local news.

WISH TV (Channels 8.1/CBS; 8.2/Local Weather Station; & 8.3/Live Doppler Radar)
WNDY TV (Channels 23.1 & 32)

Mailing Address	P.O. Box 7088, Indianapolis IN 46207	
Street Address	1950 N. Meridian St., Indianapolis IN 46202	
Telephone	317-923-8888	
Telephone (news)	317-956-8580	
Fax	317-926-1144	
Fax (news)	317-931-2242	
E-mail (news)	newsdesk@wishtv.com	
On-air Hours	24/7	
Broadcast Company	LIN Television	
General Manager	Jeff White	jwhite@wishtv.com
News Director	Patti McGettigan	PMcGettigan@wishtv.com
News Assignment Editor	Jim Scott	jscott@wishtv.com
Sports Director	Chris Widlic	cwidlic@wishtv.com
Program Director	Lance Carwile	lcarwile@wishtv.com
Public Affairs Director	Tina Cosby	tcosby@wishtv.com
Promotions Director	Scott Hainey	shainey@wishtv.com
Accepts PSAs?	yes (contact Tina Cosby)	

WISH TV

E-mail	programming@wishtv.com	
Web Site	www.wishtv.com	
Network	CBS	
General Sales Mgr.	Julie Zoumbaris	juliez@wishtv.com
Local Sales Manager	Greg Kiley	gkiley@wishtv.com

WNDY TV

Web Site	www.INDYTV.com	
Network	MyNetwork TV	
General Sales Mgr.	Marc Elliott	melliott@MyNDYtv.com

WKOG TV (Channel 31)

Mailing Address	P.O. Box 44007, Indianapolis IN 46244
Telephone	317-920-3000
E-mail	srsue@catholic-television.tv
Web Site	www.catholic-television.tv
Broadcast Company	Kingdom of God Broadcasting
General Manager	Sister Sue Jenkins

WRTV TV (Channels 6.1/ABC & 6.2/6News)

Address	1330 N. Meridian St., Indianapolis IN 46202	
Telephone	317-635-9788	
Telephone (news)	317-269-1440	
Fax	317-269-1402	
Fax (news)	317-269-1445	
Fax (sales)	317-269-1400	
E-mail (news)	6news@6news.com	
Web Site	www.theindychannel.com	
Network Affiliation	ABC	
On-air Hours	24/7	
Broadcast Company	McGraw-Hill Broadcasting	
General Manager	Don Lundy	don_lundy@wrtv.com
News Director	Sheldon Ripson	sheldon_ripson@6news.com
News Assignment Editor	Dave Brinkers	dave_brinkers@wrtv.com
Managing Editor	John Emmert	john_emmert@6news.com
Sports Director	Dave Furst	dave_furst@wrtv.com
General Sales Manager	Sally Kohn	sally_kohn@wrtv.com
Program/Promotions Dir.	Paul Montgomery	paul_montgomery@wrtv.com
Public Affairs Director	Terri Cope Walton	terri_cope@wrtv.com
Executive Producer	Brady Gibson	brady_gibson@wrtv.com
Accepts PSAs?	yes (contact Terri Cope Walton)	

WTHR TV (Channels 13.1/NBC; 13.2/SkyTrak Weather; & 13.3/Universal Sports)

Mailing Address	P.O. Box 1313, Indianapolis IN 46206	
Street Address	1000 N. Meridian St., Indianapolis IN 46204	
Telephone	317-636-1313	
Telephone (news)	317-655-5740	
Fax	317-636-3717	
Fax (news)	317-632-6720	
Fax (sales)	317-636-9846	
E-mail (news)	newsdesk@wthr.com	
Web Site	www.wthr.com	
Network Affiliation	NBC	
On-air Hours	24/7	
Broadcast Company	Dispatch Broadcast Group	
General Manager	Jim Tellus	jtellus@wthr.com
General Sales Manager	Tim Warner	twarner@wthr.com
Local Sales Manager	Jeff Teague	jteague@wthr.com
Community Affairs Dir.	Angela Cain	acain@wthr.com
Promotions Director	Jeff Dutton	jdutton@wthr.com
Accepts PSAs?	yes (contact Young-Hee Yedinak, yyedinak@wthr.com)	

WTTK TV (Channel 29) — WTTV TV (Channel 48) — WXIN TV (Channel 45)

Address	6910 Network Pl., Indianapolis IN 46278
Telephone	317-632-5900
On-air Hours	24/7
Broadcast Company	Tribune Broadcasting
General Manager	Jerry Martin
General Sales Manager	Tim McNamara
Program Director	Harry Ford
Creative Services Director	Kurt Tovey
Accepts PSAs?	yes (contact Mari Yamaguchi)

WTTK TV & WTTV TV

Fax	317-715-6251
Fax (sales)	317-715-6250
Web Site	www.indianas4.com
Network	The CW
Local Sales Manager	Monte Costes
Notes	WTTK serves Kokomo and is a simulcast of WTTV.

WXIN TV

Telephone (news)	317-687-6541
Fax	317-687-6532
Fax (news)	317-687-6556
Fax (sales)	317-687-6531
E-mail (news)	fox59news@tribune.com
Web Site	www.fox59.com
Network	Fox
News Director	Lee Rosenthal
Sports Director	Chris Hagan
Local Sales Manager	Dennis Christine

JASONVILLE

Greene County

Jasonville Independent

Mailing Address	223 S. Lawton St., Jasonville IN 47438
Street Address	113 W. Main St., Jasonville IN 47438
Telephone/Fax	812-665-2657
E-mail	jvilleindependent@att.net
Web Site	www.jasonvilleindependent.com
Publication Date	Weekly (Wednesday)
Circulation	1,000 (paid)
Publishers/Editors	Dorman & Marilyn Clark

JASPER

The Herald
Dubois County

Mailing Address	P.O. Box 31, Jasper IN 47547
Street Address	216 E. 4th St., Jasper IN 47546
Telephone	812-482-2424
Telephone (toll-free)	877-482-2424
Telephone (news)	812-482-2626
Fax	812-482-4104
Fax (news)	812-482-5241
Fax (advertising)	812-634-7142
E-mail (news)	news@dcherald.com
Web Site	www.dcherald.com
Publication Date	Daily (Monday-Saturday)
Circulation	12,600 (paid)
Publishing Company	Jasper Herald Co.
Publishers	Dan Rumbach & John Rumbach
Editor	John Rumbach jrumbach@dcherald.com
Managing Editor	Hak Haskins
Advertising Manager	Don Shreve ads@dcherald.com
Sports Editor	Jason Recker sports@dcherald.com
City Editor	Martha Rasche mrasche@dcherald.com

See **WBDC** (Huntingburg).

See **WBTO** (Vincennes).

WITZ (990 AM) — WITZ (104.7 FM) — WQKZ (98.5 FM)

Mailing Address	P.O. Box 167, Jasper IN 47547
Street Address	1978 S. WITZ Rd., Jasper IN 47546
Telephone	812-482-2131
Telephone (toll-free)	800-206-6605
Fax	812-482-9609
On-air Hours	24/7
Sales/Operations Manager	Gene Kuntz
News Director	Brandon Elliott
Accepts PSAs?	yes

WITZ AM & FM

E-mail	witzamfm@psci.net
E-mail (news)	news@witzamfm.com
Web Site	www.witzamfm.com
Wattage	1,000 (WITZ AM)
	50,000 (WITZ FM)
Format	Adult Contemporary
Networks	ABC, Cromwell Ag, Learfield, MRN, Network Indiana, PRN
Broadcast Company	Jasper On The Air, Inc.
General Manager	G. Earl Metzger
Program/Sports Dir.	Walt Ferber
Public Affairs Director	Karen Dorrell
Notes	WITZ AM & FM are simulcast.

WQKZ

E-mail	wqkz@psci.net
Wattage	6,000
Format	Country
Network	Jones
Broadcast Company	GEM Communications
General Manager	Gene Kuntz
Program Director	Gary Lee
Sports Director	Chris James

See **WJPR** (Loogootee).

WJTS TV (Channel 18)

Mailing Address	P.O. Box 1009, Jasper IN 47547	
Street Address	511 Newton St., STE 202, Jasper IN 47546	
Telephone	812-482-2727	
Telephone (toll-free)	800-504-9587	
Fax	812-482-3696	
E-mail	mailbox@wjts.tv	
E-mail (news)	news@wjts.tv	
Web Site	www.wjts.tv	
Network Affiliation	independent	
On-air Hours	24/7	
Broadcast Company	DC Broadcasting, Inc.	
General Manager	Paul Knies	pknies@dcbroadcasting.com
News/Sports Director	Kurt Gutgsell	sports@wjts.tv
News Assignment Editor	Peter Marshall	news@wjts.tv
General Sales Manager	Bill Potter	gm@dcbroadcasting.com
Local Sales Manager	Chris Lowry	lowry@wjts.tv
Program Director	Sandra Elmore	traffic@wjts.tv
Public Affairs Director	Jeremy Markos	mailbox@wjts.tv
Accepts PSAs?	yes (contact Jeremy Markos)	

JEFFERSONVILLE

Clark County

Evening News

Address	221 Spring St., Jeffersonville IN 47130
Telephone	812-283-6636
Telephone (news)	812-206-6397
Fax	812-284-7081
Fax (news)	812-206-4598
E-mail (news)	newsroom@newsandtribune.com
Web Site	www.newsandtribune.com
Publication Date	Daily (Tuesday-Sunday)
Circulation	8,000 (paid)
Publishing Company	Newspaper Holdings Inc.
Publisher	Jim Grahn	jim.grahn@newsandtribune.com
Executive Editor	Steve Kozarovich	steve.kozarovich@newsandtribune.com
Managing Editor	Shea Van Hoy	shea.vanhoy@newsandtribune.com
Advertising Manager	Mary Tuttle	mary.tuttle@newsandtribune.com
Sports Editor	Mike Hutsell	sports@newsandtribune.com

KENDALLVILLE

Noble County

See **Kendallville Mall** (Avilla).

News-Sun

Mailing Address	P.O. Box 39, Kendallville IN 46755
Street Address	102 N. Main St., Kendallville IN 46755
Telephone	260-347-0400
Fax	260-347-7281
Fax (news)	260-347-2693
Fax (advertising)	260-347-7282
Web Site	www.thenewssunonline.com
Publication Date	Daily (Sunday-Saturday)
Circulation	8,400 (paid)
Publishing Company	KPC Media Group
Publisher	Terry Housholder	terryh@kpcnews.net
Editor	Matt Getts	mattg@kpcnews.net
Sports Editor	Justin Penland	jpenland@kpcnews.net
V.P. Sales & Marketing	Bret Jacomet	bretj@kpcnews.net
TMC/Shopper	Smart Shopper (weekly)

WAWK (1140 AM)

Address	931 East Ave., Kendallville IN 46755
Telephone	260-347-2400
Fax	260-347-2524
E-mail	don@wawk.com
Web Site	www.wawk.com
Wattage	250
Format	Oldies
Network Affiliations	Brownfield, Network Indiana, USA
On-air Hours	6:00 a.m. - 10:00 p.m.
Broadcast Company	Northeast Indiana Broadcasting
General Manager	Don Moore	don@wawk.com
News/Sports Director	Mike Shultz	mike@locl.net
Program Director	Scott Pawl	scott@locl.net
Accepts PSAs?	yes (contact Scott Pawl)

KENTLAND

Newton County

Brook Reporter — Morocco Courier — Newton County Enterprise

Mailing Address	P.O. Box 107, Kentland IN 47951
Street Address	305 E. Graham St., Kentland IN 47951
Telephone	219-474-5532
Fax	219-474-5354
E-mail	editor@sugardog.com
Web Site	www.newtoncountyenterprise.com
Publication Date	Weekly (Wednesday)
Circulation	800-paid (Brook Reporter and Morocco Courier)
	1,800-paid (Newton County Enterprise)
Publishing Company	Kankakee Valley Publishing Co. (Brook Reporter and Morocco Courier)
	Twin States Publishing (Newton County Enterprise)
Publisher	Don Hurd dongo75@aol.com
Managing Editor	Cheri Glancy editor@sugardog.com
Advertising Manager	Arlaina Janowski kentlandsales@aol.com
Notes	Brook Reporter serves Brook. Morocco Courier serves Morocco. Newton County Enterprise serves Newton County.

WIVR (101.7 FM)

Mailing Address	P.O. Box 758, Bourbonnais IL 60914
Street Address	202 E. Walnut, Watseka IL 60970
Telephone	815-432-0700
Fax	815-933-8696
E-mail	wivrfm@comcast.net
Web Site	www.rivervalleyradio.net
Wattage	6,000
Format	Country
Network Affiliations	AP, Chicago Bears
On-air Hours	24/7
Broadcast Company	Milner Broadcasting
General Manager	Tim Milner tim@rivervalleyradio.net
News Director	Ken Zyer wvliwivrnews@comcast.net
Program Director	Mickey Milner mick@rivervalleyradio.net
Sales Manager	Chris Swain swain@rivervalleyradio.net
Public Affairs Director	Gordy McCollum gordy@rivervalleyradio.net
Farm Director	Mike Ruble ruble@rivervalleyradio.net
Operations Director	Jim Brandt brandt@rivervalleyradio.net
Accepts PSAs?	yes
Notes	Serves Kentland & Rensselaer, IN; Kankakee & Watseka, IL.

KEWANNA

Fulton County

Observer

Mailing Address	P.O. Box 307, Kewanna IN 46939
Street Address	110 E. Main St., Kewanna IN 46939
Telephone	574-653-2101
Fax	574-653-3418
E-mail	jkgood@rtcol.com
Publication Date	Weekly (Thursday)
Circulation	600 (paid)
Owner/Publisher/Editor	Karen Good

KNIGHTSTOWN

The Banner
Henry County

Mailing Address	P.O. Box 116, Knightstown IN 46148
Street Address	24 N. Washington St., Knightstown IN 46148
Telephone	765-345-2292
Fax	765-345-2113
E-mail	thebanner@embarqmail.com
Web Site	www.thebanneronline.com
Publication Date	Weekly (Wednesday)
Circulation	2,000 (paid)
Publisher/Editor/Ad Mgr.	Eric Cox	thebanner@embarqmail.com
Sports Editor	Ty Swincher	bannersports@embarqmail.com
Freelance Reporter	Jeff Eakins	bannerjeff@embarqmail.com

WKPW (90.7 FM)

Address	10892 N. State Rd. 140, Knightstown IN 46148
Telephone	765-345-9070
E-mail	wkpw@ktownonline.net
Web Site	www.wkpwfm.com
Wattage	4,400
Format	Classic Hits
On-air Hours	24/7
Owner	New Castle Area Career Programs
General Manager	Robert Hobbs
Program Director	Mike York	wkpw@ktownonline.net
Accepts PSAs?	yes (contact Mike York)
Notes	Non-commercial, student-operated station. Does not broadcast local news.

KNOX

The Leader of Starke County
Starke County

Street Address	15 N. Main St., Knox IN 46534
Telephone	574-772-2101
Fax	574-772-7041
E-mail	theleader@nitline.net
Publication Date	Weekly (Thursday)
Circulation	3,500 (paid)
Publishing Company	Horizon Publications
Publisher	Rick Kreps	rkreps@thepilotnews.com
Editor	John Reed	theleader@nitline.net
Managing Editor	Maggie Nixon	mnixon@thepilotnews.com
TMC/Shopper	The Review (weekly)
Notes	Main office at The Pilot-News (Plymouth).

WKVI (1520 AM) — WKVI (99.3 FM)

Mailing Address	P.O. Box 10, Knox IN 46534
Street Address	400 W. Culver Rd., Knox IN 46534
Telephone	574-772-6241
Fax	574-772-5920
E-mail (news)	news@wkvi.com
Web Site	www.wkvi.com
Wattage	250 (WKVI AM)
	3,000 (WKVI FM)
Format	Adult Contemporary
Network Affiliations	Brownfield
On-air Hours	sunrise - sunset (WKVI AM)
	24/7 (WKVI FM)
Broadcast Company	Kankakee Valley Broadcasting
General Mgr./Program Dir.	Ted Hayes spots@wkvi.com
News Director	Anita Goodan anita@wkvi.com
Sports Director	Harold Welter sports@wkvi.com
Sales/Public Affairs Mgr.	Chris Milner spots@wkvi.com
Accepts PSAs?	yes (contact Chris Milner)
Notes	WKVI AM & FM are simulcast.

KOKOMO

Howard County

Kokomo Herald

Mailing Address	P.O. Box 6488, Kokomo IN 46904
Street Address	207 N. Buckeye St., Kokomo IN 46902
Telephone	765-452-5942
Fax	765-452-3037
E-mail	editor@kokomoherald.com
Web Site	www.kokomoherald.com
Publication Date	Weekly (Thursday)
Circulation	1,100 (paid)
Publishing Company	Herald Publishing Corporation
Publisher/Executive Editor	Wayne Janner wjanner@kokomoherald.com

Kokomo Perspective

Address	209 N. Main St., Kokomo IN 46901
Telephone	765-452-0055
Fax	765-457-7209
E-mail	editor@kokomoperspective.com
Web Site	www.kokomoperspective.com
Publication Date	Weekly (Wednesday)
Circulation	31,500 (free/delivered & newsstand)
Publishing Company	Wilson Advertising
Publisher	Don Wilson don@kokomoperspective.com
Managing Editor	Lisa Fipps lfipps@kokomoperspective.com
Advertising Manager	William Eldridge eldridge@kokomoperspective.com
Sports Editor	Steve Geiselman sgeiselman@kokomoperspective.com

Kokomo Tribune

Mailing Address	P.O. Box 9014, Kokomo IN 46904
Street Address	300 N. Union St., Kokomo IN 46901
Telephone	765-459-3121
Telephone (toll-free)	800-382-0696
Telephone (news)	765-454-8584
Fax	765-456-3815
Fax (news)	765-854-6733
Fax (advertising)	765-854-6746
E-mail (news)	ktnews@kokomotribune.com
Web Site	www.kokomotribune.com
Publication Date	Daily (Sunday-Saturday)
Circulation	20,500 (paid)
Publishing Company	Community Newspaper Holdings Inc.
Publisher/Executive Editor	Robyn McCloskey robyn.mccloskey@indianamediagroup.com
Managing Editor	Jeff Kovaleski jeff.kovaleski@kokomotribune.com
Assistant Managing Editor	Misty Knisely misty.knisely@kokomotribune.com
Advertising Manager	Kristin Johnson kristin.johnson@kokomotribune.com
Sports Editor	Dave Kitchell dave.kitchell@kokomotribune.com
Business Editor	K. O. Jackson kirven.jackson@kokomotribune.com
TMC/Shopper	Weekly

See **WFIU** (Bloomington).

See **WFRR** (Elkhart).

WIOU (1350 AM) — WMYK (98.5 FM) — WZWZ (92.5 FM)

Mailing Address	P.O. Box 2208, Kokomo IN 46904
Street Address	671 E. 400 S., Kokomo IN 46902
Fax	765-455-3882
E-mail (news)	newsroom@z925fm.com
On-air Hours	24/7
Broadcast Company	Hoosier AM/FM LLC
General Manager	Steve La Mar svlamar76@aol.com
News Director	Barb Wire newsroom@z925fm.com
Program Director	Allan James allanjames925@aol.com
Sports Director	Greg Bell greg@z925fm.com
Sales Manager	Lora Lacy lora@z925fm.com
Accepts PSAs?	yes (contact Beth Dillon, radioreceptionist@sbcglobal.net)

WIOU

Telephone	765-453-1212
E-mail	wzwz-wiou-wmyk@sbcglobal.net
Wattage	5,000
Format	Talk/News/Sports
Network	CBS
Promotions Director	Allan James allanjames925@aol.com

WMYK

Telephone	765-455-9850
Phone (toll-free)	866-985-7625
E-mail	classicrock985@sbcglobal.net
Web Site	www.rock985.com
Wattage	6,000
Format	Rock
Networks	Free Beer and Hot Wings, Alice Cooper
Promotions Director	Mike Turner mike@z925fm.com

WZWZ

Telephone	765-453-1212
Phone (toll-free)	866-992-7625
E-mail	wzwz-wiou-wmyk@sbcglobal.net
Web Site	www.z925fm.com
Wattage	6,000
Format	Adult Contemporary
Network	ABC
Promotions Director	Allan James allanjames925@aol.com

See **WIWC** (Anderson).

WJJD (101.3 FM)

Address	1763 E. 100 N., Kokomo IN 46901
Telephone	765-452-3300
Wattage	100
Format	Religious
Network Affiliations	Radio 74
On-air Hours	24/7
Broadcast Company	Kokomo Seventh Day Adventist Broadcasting
General Manager	Blake Hall blakerh@hotmail.com
Accepts PSAs?	yes

WWKI (100.5 FM)

Address	519 N. Main St., Kokomo IN 46901
Telephone	765-459-4191
Telephone (toll-free)	800-456-1106
Telephone (news)	765-456-1104
Fax	765-456-1111
Fax (news)	765-456-1112
E-mail	wwki@citcomm.com
E-mail (news)	dave.broman@citcomm.com
Web Site	www.wwki.com
Wattage	50,000
Format	Hit Country
On-air Hours	24/7
Broadcast Company	Citadel Broadcasting Co.
General Manager	Mike Christopher mike.christopher@citcomm.com
News/Program Director	Dave Broman dave.broman@citcomm.com
Sales Manager	Jim Stonecipher jim.stonecipher@citcomm.com
Accepts PSAs?	yes (contact David Broman)

See **WTTK TV** (Indianapolis).

LAFAYETTE

See **The Catholic Moment** (listed under Religion/Specialty Publications).   Tippecanoe County

See **HELEN Magazine** (listed under Women/Specialty Publications).

Journal and Courier

Address	217 N. 6th St., Lafayette IN 47901
Telephone	765-423-5511
Telephone (toll-free)	800-456-3223
Telephone (news)	765-420-5259
Fax (news)	765-420-5246
Fax (advertising)	765-742-5633
E-mail (news)	dbangert@journalandcourier.com
Web Site	www.jconline.com
Publication Date	Daily (Sunday-Saturday)
Circulation	36,000-paid (daily); 46,000-paid (Sunday)
Publishing Company	Gannett Inc.
Publisher	Gary Suisman gsuisman@journalandcourier.com
Executive Editor	Julie Doll jdoll@journalandcourier.com
Managing Editor	Henry Howard hhoward@journalandcourier.com
Advertising Manager	Jim Holm jholm@journalandcourier.com
Sports Editor	Jim Stafford jstafford@journalandcourier.com
City Editor	Dave Bangert dbangert@journalandcourier.com
TMC/Shopper	J & C Weekly
Notes	Also Publishes **en espanol** (monthly issue of Journal and Courier) in Spanish.

Lafayette Leader

Address	401 Main St., STE 2F, Lafayette IN 47901
Telephone	765-428-8123
Fax	765-428-8124
E-mail	business@lafayetteleader.net
Publication Date	Weekly (Thursday)
Circulation	2,000 (paid)
Publishing Company	Community Media Group
President	Don Hurd dongo75@aol.com
Editor	Scott Allen sallen@thehj.com

WASK (1450 AM) — WASK (98.7 FM) — WKHY (93.5 FM) — WKOA (105.3 FM) WXXB (102.9 FM)

Address	3575 McCarty Lane, Lafayette IN 47905
On-air Hours	24/7
Broadcast Company	Schurz Communications
General Manager	John Schurz jschurz@wask.com
Director of Sales	Brian Green bgreen@wask.com
Promotions Director	Liz Hahn lhahn@wask.com
Notes	Local news provided by WLFI TV.

WASK AM

Telephone	765-447-2186
Fax	765-448-4452
Wattage	1,000
Format	ESPN Sports
Program Director	Bryan McGarvey mcgarvey@wask.com
Accepts PSAs?	yes (contact Bryan McGarvey)

WASK FM

Telephone	765-447-2186
Fax	765-448-4452
Web Site	www.wask.com
Wattage	4,000
Format	Super Hits 60's & 70's
Program Director	Bryan McGarvey mcgarvey@wask.com
Accepts PSAs?	yes-only pre-recorded PSAs (contact Bryan McGarvey)

WKHY

Telephone	765-448-1566
Fax	765-448-1348
Web Site	www.wkhy.com
Wattage	6,000
Format	Rock
Program Director	Jeff Strange strange@wkhy.com
Accepts PSAs?	yes (contact Jeff Strange)

WKOA

Telephone	765-447-2186
Fax	765-448-4452
Web Site	www.wkoa.com
Wattage	50,000
Format	Country
Program Director	Mike Shamus shamus@wkoa.com
Accepts PSAs?	yes (contact Mike Shamus)

WXXB

Telephone	765-448-1566
Fax	765-448-1348
Web Site	www.b1029.com
Wattage	6,000
Format	Contemporary Hit Radio Mainstream
Program Director	Anthony Bannon tonyb@b1029.com
Accepts PSAs?	yes

WAZY (96.5 FM) — WBPE (95.3 FM) — WSHP (95.7 FM) — WSHY (1410 AM)

Address	3824 S. 18th St., Lafayette IN 47909
Telephone	765-474-1410
Fax	765-474-3442
On-air Hours	24/7
Broadcast Company	Artistic Media Partners
General Manager	Ernie Caldemone ernie@artisticradio.com
Accepts PSAs?	yes
Notes	Does not broadcast local news.

WAZY

Web Site	www.wazy.com
Wattage	50,000
Format	Top 40/Mainstream
Program Director	Jimmy Knight jimmy@wazy.com
Sales Manager	Kit Osborne kit@wazy.com

WBPE & WSHY

Web Site	www.wbpefm.com
Wattage	6,000 (WBPE)
	2,300 (WSHY)
Format	Adult Hits/BOB-FM
Program Director	Jimmy Knight jimmy@wazy.com
Sales Manager	Kit Osborne kit@wazy.com
Notes	WBPE & WSHY are simulcast.

WSHP

Web Site	www.957therocket.com
Wattage	6,000
Format	Classic Rock
Program Director	Rob Creighton rob@957therocket.com
Sales Manager	Fred Stuart fred@957therocket.com

See **WHPL** (Anderson).

WJEF (91.9 FM)

Address	1801 S. 18th St., Lafayette IN 47905
Telephone	765-772-4700
E-mail	rbrist@lsc.k12.in.us
Web Site	www.jeff92.org
Wattage	250
Format	Oldies
On-air Hours	24/7
Owner	Lafayette School Corp.
General Manager	Randy Brist rbrist@lsc.k12.in.us
Accepts PSAs?	yes (contact Randy Brist)
Notes	Non-commercial station. Broadcasts local news only during school year.

See **WKHL/K-Love** (listed under National Radio Stations).

See **WQSG/American Family Radio** (listed under National Radio Stations).

WTGO (97.7 FM)

Address	724 Wabash Ave., Lafayette IN 47905
Telephone	765-429-1113
E-mail	rockshowdj@gmail.com
Web Site	www.wtgoradio.com
Wattage	100
Format	Christian Rock
Owner	Harvest Chapel
General Manager	Tom Camp tomcamp50@gmail.com
Program Director	Brett Estes rockshowdj@gmail.com

WWCC (97.3 FM)

Address	101 N. 10th St., Lafayette IN 47901
Telephone	765-474-3776
Fax	765-423-2343
E-mail	info@wwcconline.org
Web Site	www.wwcconline.org
Wattage	14
Format	Christian
On-air Hours	24/7
Owner	Triangle Foundation
General Manager	Bob Schueler
Notes	Does not broadcast local news.

LaGRANGE

LaGrange County

LaGrange News — LaGrange Standard — Middlebury Independent

Street Address	State Rd. 9 & C.R. 100 S., LaGrange IN 46761
Telephone	260-463-2166
Telephone (toll-free)	800-552-2404
Fax	260-463-2734
E-mail	lagpubco@kuntrynet.com
Web Site	www.lagrangepublishing.com
Publishing Company	LaGrange Publishing Co.
Publisher	William Connelly
Editor	Guy Thompson
Advertising Manager	Scott Faust
Sports Editor	Jeff Miller

LaGrange News

Mailing Address	P.O. Box 148, LaGrange IN 46761
Publication Date	Weekly (Friday)
Circulation	5,200 (paid)
TMC/Shopper	The Countian (weekly)

LaGrange Standard

Mailing Address	P.O. Box 148, LaGrange IN 46761
Publication Date	Weekly (Monday)
Circulation	5,200 (paid)
TMC/Shopper	The Countian (weekly)

Middlebury Independent

Mailing Address	P.O. Box 68, Middlebury IN 46540
Telephone	574-825-9112
Publication Date	Weekly (Wednesday)
Circulation	1,000 (paid)
TMC/Shopper	Crystal Valley Trading Post (weekly)
Notes	Serves Middlebury.

WTHD (105.5 FM)

Address	206 S. High St., LaGrange IN 46761
Telephone	260-463-8500
Fax	260-463-8580
E-mail	wthd@wthd.net
Web Site	www.wthd.net
Wattage	6,000
Format	Country
Network Affiliations	ABC, Jones
On-air Hours	24/7
Broadcast Company	Swick Broadcasting
General Mgr./Sales Mgr.	Steve Swick
News Director	Tim Murray tmurray@wthd.net
Accepts PSAs?	yes

LAPEL

See **Times/Post** (Pendleton).

Madison County

LaPORTE

LaPorte County

Herald-Argus

Address	701 State St., LaPorte IN 46350
Telephone	219-362-2161
Telephone (toll-free)	866-362-2167
Fax	219-362-2166
E-mail	ha@heraldargus.com
Web Site	www.heraldargus.com
Publication Date	Daily (Monday-Saturday)
Circulation	11,000 (paid)
Publishing Company	Paxton Media Group
Publisher	Patrick Kellar pkellar@heraldargus.com
Executive Editor	Chris Schable cschable@heraldargus.com
Advertising Manager	Brad Reisig breisig@heraldargus.com
Sports Editor	Adam Parkhouse aparkhouse@thenewsdispatch.com
TMC/Shopper	Herald News Review (weekly)

WCOE (96.7 FM) — WLOI (1540 AM)

Address	1700 Lincolnway Pl., STE 8, LaPorte IN 46350
Telephone (news)	219-362-5515
Fax	219-324-7418
E-mail (news)	newshawk7@yahoo.com
Web Site	www.wcoefm.com
Network Affiliations	CNN
Broadcast Company	LaPorte County Broadcasting
General Manager	Dennis Siddall denny@wcoefm.com
News Director	Stan Maddux newshawk7@yahoo.com
Sales Manager	Norma Sabie normas@wcoefm.com
Accepts PSAs?	yes

WCOE

Telephone	219-362-5290
Wattage	3,000
Format	Country
On-air Hours	24/7

WLOI

Telephone	219-362-6144
Wattage	250
Format	Adult Standards
On-air Hours	daytime

LAWRENCEBURG

Dearborn County

Dearborn County Register

Mailing Address	P.O. Box 4128, Lawrenceburg IN 47025
Street Address	126 W. High St., Lawrenceburg IN 47025
Telephone	812-537-0063
Fax	812-537-5576
E-mail	regpublisher@gmail.com
Web Site	www.thedcregister.com
Publication Date	Weekly (Thursday)
Circulation	7,400 (paid)
Publishing Company	Register Publications
Publisher	Joe Awad	editor@registerpublications.com
Editor	Erika Russell	community@registerpublications.com
Advertising Manager	Loretta Day	lday@registerpublications.com
Sports Editor	Jim Buchberger	sports@registerpublications.com
TMC/Shopper	Market Place (weekly)

See **Over Fifty Magazine** (listed under Senior Citizens/Specialty Publications).

WSCH (99.3 FM)

Address	20 E. High St., Lawrenceburg IN 47025
Telephone	812-537-0944
Telephone (toll-free)	888-537-9724
Fax	812-537-5735
E-mail	info@eaglecountryonline.com
E-mail (news)	news@eaglecountryonline.com
Web Site	www.eaglecountryonline.com
Wattage	3,000
Format	Top 40 Country
Network Affiliations	ABC, Learfield, MRN, PRN
On-air Hours	24/7
Broadcast Company	Wagon Wheel Broadcasting LLC
General Manager	Mike Shields
Program Director	Chelsie Shinkel
Accepts PSAs?	yes (contact Chelsie Shinkel)

LEBANON

Boone County

Lebanon Reporter

Address	117 E. Washington St., Lebanon IN 46052
Telephone	765-482-4650
Telephone (toll-free)	888-482-4650
Fax	765-482-4652
E-mail (news)	news@reporter.net
Web Site	www.reporter.net
Publication Date	Daily (Monday-Saturday)
Circulation	5,200 (paid)
Publishing Company	CNHI
Publisher	Greta Sanderson	greta.sanderson@reporter.net
Managing Editor	Marda Johnson	marda.johnson@reporter.net
Advertising Manager	Rick Whiteman	rick.whiteman@reporter.net
Sports Editor	Will Willems	will.willems@reporter.net
TMC/Shopper	weekly

WIRE (91.1 FM)

Address	3500 DePauw Blvd., STE 2085, Indianapolis IN 46268
Telephone	317-870-8400
Telephone (toll-free)	866-883-8400
Fax	317-870-8404
E-mail (news)	scott@radiomom.fm
Web Site	www.radiomom.fm
Wattage	3,000
Format	Adult Contemporary
On-air Hours	24/7
Broadcast Company	Hoosier Broadcasting Corp.
General Manager	Chuck Cunningham cac@hoosierbroadcastingcorp.org
News/Sports Director	Scott Carney scott@radiomom.fm
Program Director	Bill Shirk billshirk@radiomom.fm
Accepts PSAs?	yes (contact Scott Carney)
Notes	Serves Lebanon. Non-commercial station.

LIBERTY

Liberty Herald — Union County Review

Union County

Mailing Address	P.O. Box 10, Liberty IN 47353
Street Address	10-12 N. Market St., Liberty IN 47353
Telephone	765-458-5114
Fax	765-458-5115
Web Site	www.libertyherald.net
Publication Date	Weekly: Thursday (Liberty Herald)
	Weekly: Tuesday (Union County Review)
Circulation	2,700-paid (Liberty Herald)
	5,000-free/mailed (Union County Review)
Publishing Company	Whitewater Publications
Publisher	Gary Wolf gary@thebrookvillenews.com
Editor	John Estridge john@thebrookvillenews.com
Advertising Manager	Melissa Lilly ads@thebrookvillenews.com
Sports Editor	Vivian Risch

LIGONIER

The Advance Leader

Noble County

Mailing Address	P.O. Box 30, Ligonier IN 46767
Street Address	102 N. Main St., Kendallville IN 46755
Telephone/Fax	260-894-3102
Web Site	www.advanceleaderonline.com
Publication Date	Weekly (Thursday)
Circulation	1,000 (paid)
Publishing Company	KPC Media Group
Publisher	Terry Housholder terryh@kpcnews.net
Editor/Advertising Mgr.	Bob Buttgen leader@kpcnews.net
Notes	KPC Media Group main office is in Kendallville.

WNRL (105.9 FM)

Address	5094 N. U.S. Highway 33, Ligonier IN 46767
Telephone	260-894-9777
E-mail	WNRL1059@Ligtel.com
Web Site	www.mix106online.com
Wattage	60
Format	Community Radio
On-air Hours	24/7
Owner	West Noble High School
General Manager	Steve Weaver wnrl@westnoble.k12.in.us
Accepts PSAs?	yes (contact Steve Weaver)

LINTON

Greene County

Greene County Daily World

Mailing Address	P.O. Box 129, Linton IN 47441
Street Address	79 S. Main St., Linton IN 47441
Telephone	812-847-4487
Telephone (toll-free)	800-947-4487
Fax	812-847-9513
E-mail	cpruett@gcdailyworld.com
Web Site	www.gcdailyworld.com
Publication Date	Daily (Tuesday-Saturday)
Circulation	4,800 (paid)
Publishing Company	Rust Publishing
Publisher	Randy List rlist2@hotmail.com
General Manager/Editor	Chris Pruett cpruett79@hotmail.com
Advertising Manager	Christy Lehman christy_lehman@hotmail.com
Sports Editor	B. J. Hargis hargisbj@yahoo.com
TMC/Shopper	The Shopper (weekly)
Notes	Serves Linton & Bloomfield.

See **WQTY** (Vincennes).

WYTJ (89.3 FM)

Address	R. R. 3, Box 1034, Linton IN 47441
Telephone	812-847-7442
Fax	812-847-7222
E-mail	wytj893fm@minerbroadband.com
Web Site	www.wytj893fm.org
Wattage	1,000
Format	Religious
Network Affiliations	FBN
On-air Hours	24/7
Owner	Bethel Baptist Church
General Manager	Harold Smith
Program Director	Kimberlie Smith
Accepts PSAs?	yes-only local PSAs (PSAs must be submitted in writing on letterhead.)
Notes	Non-commercial station. Does not broadcast local news.

LOGANSPORT

Cass County

Info: Logansport's Bilingual Newspaper

Mailing Address	P.O. Box 314, Logansport IN 46947
Street Address	206 Fourth St., Logansport IN 46947
Telephone	574-727-0396
Telephone (advertising)	574-727-0613
Fax	574-735-0559
E-mail (news)	info@existentialmedia.com
Publication Date	Monthly
Circulation	5,500 (free)
Publishing Company	Existential Media
Publisher/Editor	Michelle Laird michelle@existentialmedia.com
Advertising Manager	Phill Dials phill@existentialmedia.com
Notes	Bilingual newspaper serving the Hispanic community.

Pharos-Tribune

Mailing Address	P.O. Box 210, Logansport IN 46947
Street Address	517 E. Broadway, Logansport IN 46947
Telephone	574-722-5000
Telephone (toll-free)	800-676-4125
Fax	574-732-5055
Fax (news)	574-732-5070
Fax (advertising)	574-732-5050
E-mail (news)	ptnews@pharostribune.com
Web Site	www.pharostribune.com
Publication Date	Daily (Sunday-Saturday)
Circulation	10,500 (paid)
Publishing Company	cnhi Media
Publisher	Robyn McCloskey robyn.mccloskey@indianamediagroup.com
General Manager	Kim Dillon kim.dillon@indianamediagroup.com
Managing Editor	Kelly Hawes kelly.hawes@pharostribune.com
Associate Editor	John Dempsey john.dempsey@pharostribune.com
Advertising Manager	Chris Ford chris.ford@pharostribune.com
Sports Editor	Beau Wicker beau.wicker@pharostribune.com
TMC/Shopper	Smart Shopper (weekly)

WHZR (103.7 FM) — WLHM (102.3 FM) — WSAL (1230 AM)

Street Address	425 2nd St., Logansport IN 46947
Fax	574-739-1037
E-mail (news)	wsalnewsroom@gmail.com
On-air Hours	24/7
Broadcast Company	Mid-America Radio Group, Inc.
General Mgr./Sales Mgr.	Daniel Keister dankeister@verizon.net
Operations Director	Bob Ehle Jr. bobehle@gmail.com
Accepts PSAs?	yes (contact Rob Morter, wsalnewsroom@gmail.com)

WHZR

Mailing Address	P.O. Box 103, Logansport IN 46947
Telephone	574-732-1037
E-mail	whzr@verizon.net
Web Site	www.hoosier1037.com
Wattage	6,000
Format	Country
Network	ABC Country Coast to Coast
Program Director	Dale Lowe dale@hoosiercountry1037.com
Sports Director	Milt Hess

WLHM

Mailing Address	P.O. Box 719, Logansport IN 46947
Telephone	574-722-4000
Web Site	www.mix102radio.com
Wattage	3,000
Format	Hot Adult Contemporary
Sports Director	Mike Montgomery

WSAL

Mailing Address	P.O. Box 719, Logansport IN 46947
Telephone	574-722-4000
Web Site	www.1230wsal.com
Wattage	1,000
Format	Full Service
Networks	Fox News, Hoosier Ag Today, Network Indiana
Sports Director	Mike Montgomery

LOOGOOTEE

Martin County

Loogootee Tribune

Mailing Address	P.O. Box 277, Loogootee IN 47553
Street Address	514 N. JFK Ave., Loogootee IN 47553
Telephone	812-295-2500
Fax	812-295-5221
E-mail	news@loogtribune.com
Publication Date	Weekly (Thursday)
Circulation	3,500 (paid)
Publishing Company	Hembree Communications
Publisher/Editor	Larry Hembree larry@loogtribune.com
Advertising Manager	Alan Williams alan@loogtribune.com

See **WBHW** (Evansville).

WJPR (91.7 FM)

Street Address	514 JFK Ave., Loogootee IN 47553
Telephone	812-295-2500
Fax	812-295-5221
Wattage	2,600
Format	Oldies
On-air Hours	24/7
Broadcast Company	Jasper Public Radio Inc.
General Manager	Larry Hembree larry@loogtribune.com
Program/Public Affairs Dir.	Alan Williams alan@loogtribune.com
Accepts PSAs?	yes (contact Alan Williams)
Notes	Non-profit station. Serves Jasper

WRZR (94.5 FM)

Address	514 JFK Ave., Loogootee IN 47553
Telephone	812-295-9480
Telephone (toll-free)	888-253-1845
Fax	812-482-3696
Fax (news)	812-295-3295
Fax (sales)	812-683-5891
E-mail	mailbox@wrzr.us
Web Site	www.wrzr.us
Wattage	3,000
Format	Classic Rock
Network Affiliations	Brownfield Ag, Colts, Dial Global, IU Sports, Network Indiana
On-air Hours	24/7
Broadcast Company	DC Broadcasting
General Manager	Bill Potter gm@dcbroadcasting.com
News Director	Mike Carie news@wrzr.us
Sports Director	Greg Bateman
Sales Manager	Ron Spaulding sales@dcbroadcasting.com
Accepts PSAs?	yes

LOWELL

Lake County

Cedar Lake Journal — Lowell Tribune

Mailing Address	P.O. Box 248, Lowell IN 46356
Street Address	116 Clark St., Lowell IN 46356
Telephone	219-696-7711
Fax (news)	219-696-7713
Fax (advertising)	219-696-7266
E-mail	pilcherpubco@comcast.net
E-mail (news)	lowelltrib@comcast.net
Publication Date	Weekly (Tuesday)
Circulation	1,400-paid (Cedar Lake Journal)
	4,900-paid (Lowell Tribune)
Publishing Company	Pilcher Publishing Co.
Publisher/Advertising Mgr.	Gary Pilcher
Editor (Cedar Lake Journal)	Louise Roys
Editor (Lowell Tribune)	Connie Schrombeck
Notes	Cedar Lake Journal serves Cedar Lake.

See **Cedar Lake/Lowell Star** (Crown Point).

See **WTMK** (Valparaiso).

MADISON

Jefferson County

Madison Courier

Address	310 Courier Square, Madison IN 47250
Telephone	812-265-3641
Telephone (toll-free)	800-333-2885
Fax	812-273-6903
E-mail (news)	news@madisoncourier.com
Web Site	www.madisoncourier.com
Publication Date	Daily (Monday-Saturday)
Circulation	9,200 (paid)
Publishing Company	Madison Courier Inc.
Publisher	Jane Jacobs	jwjacobs@madisoncourier.com
Editor	Elliot Tompkin	etompkin@madisoncourier.com
Advertising Manager	Mark McKee	mmckee@madisoncourier.com
Sports Editor	Mark Campbell	mcampbell@madisoncourier.com
TMC/Shopper	Weekly Herald (weekly)

RoundAbout Entertainment Guide

Address	314 Jefferson St., STE 201, Madison IN 47250
Telephone	812-273-2259
Telephone (toll-free)	800-343-3005
Fax	812-273-5490
Fax (news)	800-277-6318
E-mail	info@roundabout.bz
Web Site	www.roundabout.bz
Publication Date	Monthly
Circulation	36,000 (free/delivered & mailed)
Publishing Company	Kentuckiana Publishing
Publisher/Editor	Donald L. Ward
Notes	Arts & entertainment publication with a regional focus. Publishes 2 editions: Indiana and Kentucky.

WIKI (95.3 FM)

Address	2604 Michigan Rd., Madison IN 47250
Telephone	812-273-3139
Telephone (toll-free)	800-953-WIKI
Fax	812-265-4536
E-mail	wiki_953@yahoo.com
Web Site	www.wikiradiofm.com
Wattage	3,000
Format	Country
On-air Hours	24/7
Broadcast Company	Wagon Wheel Broadcasting LLC
General Manager	Mike Shields	mikeshields@eaglecountryonline.com
News/Program Director	Larry Duke	wiki_953@yahoo.com
Sales Manager	Ernie Ernstes	maernstes@hotmail.com
Accepts PSAs?	yes (contact Larry Duke)

WORX (96.7 FM) — **WXGO** (1270 AM)

Mailing Address	P.O. Box 95, Madison IN 47250
Street Address	1224 E. Telegraph Hill Rd., Madison IN 47250
Telephone	812-265-3322
Telephone (toll-free)	800-660-9679
Fax	812-273-5509
E-mail	thebestmusic@worxradio.com
E-mail (news)	newsports@worxradio.com
Web Site	www.worxradio.com
Wattage	3,000 (WORX)
	1,000 (WXGO)
Format	Hot Adult Contemporary (WORX)
	Oldies (WXGO)
Network Affiliations	Jones
On-air Hours	24/7
Broadcast Company	DC Broadcasting
General Manager	Bill Potter gm@dcbroadcasting.com
News/Sports Director	Casey Bloos newsports@worxradio.com
Operations Manager	Tim Torrance timmyt@worxradio.com
Accepts PSAs?	yes (contact Tim Torrance)

MARENGO

Crawford County

WBRO (89.9 FM)

Mailing Address	P.O. Box 181, Marengo IN 47140
Street Address	1130 S. S.R. 66, Marengo IN 47140
Telephone	812-365-9276
Fax	812-365-2127
E-mail	wbrofm@aol.com
Web Site	www.wbro.org
Wattage	1,000
Format	Variety
Network Affiliations	Learfield, Network Indiana, PRI
On-air Hours	24/7
Broadcast Company	Southern Indiana Community Communications
General Manager	Shawn Scott
Accepts PSAs?	yes (contact Shawn Scott)
Notes	Non-commercial station.

MARION

Grant County

Chronicle-Tribune

Mailing Address	P.O. Box 309, Marion IN 46952
Street Address	610 S. Adams St., Marion IN 46953
Telephone	765-664-5111
Telephone (toll-free)	800-955-7888
Telephone (news)	765-671-1266
Fax	765-664-6292
Fax (news)	765-668-4256
Fax (advertising)	765-664-0729
E-mail	reporter@comteck.com
Web Site	www.chronicle-tribune.com
Publication Date	Daily (Monday-Saturday)
Circulation	18,000 (paid)
Publishing Company	Paxton Media
Publisher	Neal Ronquist nronquist@paxtonmedia.com
Managing Editor	David Penticuff dpenticuff@chronicle-tribune.com
Sports Editor	Justin Kenny jkenny@chronicle-tribune.com
TMC/Shopper	The Extra (weekly)

See **The Giant** (Gas City).

See **Sports Hotline** (listed under Sports/Specialty Publications).

WBAT (1400 AM) — WCJC (99.3 FM) — WMRI (860 AM) — WXXC (106.9 FM)

Mailing Address	P.O. Box 839, Marion IN 46952
Street Address	820 S. Pennsylvania St., Marion IN 46953
E-mail (news)	news@wbat.com
On-air Hours	24/7
Broadcast Company	Hoosier AM/FM LLC
General Manager	David Poehler	dpoehler@comteck.com
News Director	Kristopher Lee	kristopher@1609wxxc.com
Sports Director	Jim Brunner	jimmyb@comteck.com

WBAT

Telephone	765-664-6239
Phone (toll-free)	866-993-9252
Fax	765-662-0730
E-mail	wbat@comteck.com
Web Site	www.wbat.com
Wattage	1,000
Format	Oldies/News/Sports
Networks	ABC, CBS, ESPN, Network Indiana, Westwood One
Program Director	Tim George	wbat@comteck.com
Sales Manager	Jim Brunner	jimmyb@comteck.com
Accepts PSAs?	yes (contact Tim George)

WCJC

Telephone	765-664-6239
Phone (toll-free)	866-993-9252
Fax	765-662-0730
E-mail	wcjc@comteck.com
Web Site	www.wcjc.com
Wattage	3,000
Format	Country
Networks	ABC, CBS, ESPN, Network Indiana, Westwood One
Program Director	Tim George	wbat@comteck.com
Sales Manager	Jim Brunner	jimmyb@comteck.com
Accepts PSAs?	yes (contact Tim George)

WMRI

Telephone	765-664-7396
Phone (toll-free)	800-765-1069
Fax	765-668-6767
E-mail	wmri@comteck.com
Web Site	www.wmri.com
Wattage	1,000
Format	Gospel
Network	Fox Radio News
Program Director.	Vanessa Miller	vanessa@wmri.com
Sales Manager	Gloria Millspaugh	gmillspaugh@wmri.com
Accepts PSAs?	yes (contact Vanessa Miller)

WXXC

Telephone	765-664-7396
Phone (toll-free)	800-765-1069
Fax	765-668-6767
E-mail	studio@1069wxxc.com
Web Site	www.1069wxxc.com
Wattage	50,000
Format	Classic Hits
Program Director.	Vanessa Miller	vanessa@wmri.com
Sales Manager	Gloria Millspaugh	gmillspaugh@wmri.com
Accepts PSAs?	yes (contact Vanessa Miller)

See **WBSW** (Muncie).

See **WFRN** (Elkhart).

WIWU (94.3 FM)

Address	4201 S. Washington St., Marion IN 46953
Telephone	765-677-2773
Fax	765-677-1042
E-mail	94.3@indwes.edu
Web Site	www.indwes.edu/thefortress
Wattage	100
Network Affiliations	CHRSN, CNN
On-air Hours	24/7
Owner	Indiana Wesleyan University
Faculty Advisor	Mark Perry
Accepts PSAs?	yes (contact 94.3@indwes.edu)

WIWU TV (Channel 51.1)

Address	4201 S. Washington St., Marion IN 46953	
Telephone	765-677-2775	
Fax	765-677-1755	
E-mail	wiwutv@indwes.edu	
Web Site	www.wiwutv51.com	
Network Affiliation	independent	
On-air Hours	24/7	
Owner	Indiana Wesleyan University	
General Manager	Paul Crisp	
Public Affairs/Promotions	Kyle Hufford	kyle.hufford@indwes.edu
Accepts PSAs?	yes (contact Kyle Hufford)	
Notes	Does not broadcast local news.	

WSOT TV (Channel 27)

Address	2172 Chapel Pike, Marion IN 46952	
Telephone	765-668-1014	
Fax	765-671-2151	
E-mail	traffic@wsot-tv.com	
Web Site	www.wsot-tv.com	
Network Affiliations	Family Net, TLN	
On-air Hours	24/7	
Broadcast Company	Sunnycrest Media, Inc.	
General Mgr./Sales Mgr.	Angela Stepp	sales@wsot-tv.com
Program Director	Floyd Broegman	traffic@wsot-tv.com
Accepts PSAs?	yes (contact Floyd Broegman)	
Notes	Does not broadcast local news.	

MARTINSVILLE

Morgan County

Reporter-Times

Mailing Address	P.O. Box 1636, Martinsville IN 46151
Street Address	60 S. Jefferson St., Martinsville IN 46151
Telephone	765-342-3311
Fax (news)	765-342-1446
Fax (advertising)	765-342-1459
E-mail (news)	bculp@reportert.com
Web Site	www.reporter-times.com
Publication Date	Daily (Sunday-Saturday)
Circulation	5,000 (paid)
Publishing Company	Schurz Communications
Publisher	E. Mayer Maloney Jr.	mmaloney@reportert.com
Managing Editor	Brian Culp	bculp@reportert.com
Advertising Manager	Cory Bollinger	cbollinger@reportert.com
City/Business Editor	Ron Hawkins	rhawkins@reportert.com
Sports Reporter	Ross Flint	rflint@reportert.com

WCBK (102.3 FM) — WMYJ (1540 AM)

Mailing Address	P.O. Box 1577, Martinsville IN 46151
Street Address	1639 Burton Lane, Martinsville IN 46151
Telephone	765-342-3394
Fax	765-342-5020
Broadcast Company	Mid-America Radio Group, Inc.
President	Dave Keister	mid-americaradio@scican.net
General Mgr./Sales Mgr.	Ruth Ann Arney	ruthann@wcbk.com
News Director	Pam Wooten	pam@wcbk.com
Program Director	John Taylor	jtaylor@wcbk.com
Sports Director	Randall Wayne Rinehart	randall@wcbk.com
Accepts PSAs?	yes (contact Steve Vail, stevevail@wcbk.com)

WCBK

E-mail	wcbk@scican.net
E-mail (news)	news@wcbk.com
Web Site	www.wcbk.com
Wattage	6,000
Format	Country
Network	ABC
On-air Hours	24/7

WMYJ

E-mail	myjoy@wcbk.com
Wattage	500
Format	Gospel
Network	Salem (Solid Gospel)
On-air Hours	sunrise - sunset
Notes	Repeats on 94.1 FM (Martinsville).

WREP TV (Channel 15.1)

Address	1360 E. Gray St., Martinsville IN 46151
Telephone	765-342-5571 (school)
Web Site	www.mhsrewind.com
Network Affiliation	America One
On-air Hours	24/7
Owner	Martinsville High School
Station Manager	Eric Meyer meyere@msdmail.net
Accepts PSAs?	yes

McCORDSVILLE
Hancock County

See **Fortville-McCordsville Reporter** (Greenfield).

MERRILLVILLE
Lake County

See **Northwest Indiana Catholic** (listed under Religion/Specialty Publications).

Post-Tribune

Address	1433 E. 83rd Ave., Merrillville IN 46410
Telephone	219-648-3000
Telephone (toll-free)	800-753-5533
Telephone (news)	219-648-3100
Fax	219-648-3249
Fax (advertising)	219-648-2187
E-mail (news)	dhayes@post-trib.com
Web Site	www.post-trib.com
Publication Date	Daily (Sunday-Saturday)
Circulation	61,200-paid (daily); 63,600-paid (Sunday)
Publishing Company	Sun-Times News Group
Publisher	Lisa Tatina ltatina@post-trib.com
Executive Editor	Paulette Haddix phaddix@post-trib.com
Managing Editor	Diane Hayes dhayes@post-trib.com
Advertising Manager	Rich Cains rcains@post-trib.com
Sports Editor	Mark Lazerus mlazerus@post-trib.com
TMC/Shopper	Crown Point Shopping News

WLPR (89.1 FM)

Address	8625 Indiana Pl., Merrillville IN 46410
Telephone	219-756-5656
Telephone (toll-free)	888-694-LAKE
Fax	219-755-4312
E-mail	info@lakeshorePTV.com
E-mail (news)	news@lakeshorePTV.com
Web Site	www.thelakeshorefm.com
Wattage	2,400
Format	News/Talk/Information
Network Affiliations	NPR
On-air Hours	24/7
Broadcast Company	Northwest Indiana Public Broadcasting
President/CEO	Tom Carroll tcarroll@lakeshorePTV.com
News Director	Ryan Priest rpriest@lakeshorePTV.com
Program Director	Len Clark lclark@lakeshorePTV.com
Sales Manager	Renee Golas rgolas@lakeshorePTV.com
Accepts PSAs?	yes (contact Kathleen Szot, kszot@lakeshorePTV.com)

WLTH (1370 AM)

Address	1563 E. 85th Ave., Merrillville IN 46410
Telephone	219-794-1370
Fax	219-794-1377
E-mail (news)	wlth1370am@sbcglobal.net
Web Site	www.wlth1370.com
Wattage	1,000
Format	Talk/News/Gospel
On-air Hours	24/7
General Mgr./Sales Mgr.	Verlie Harris
Station Manager	Mary Harris
Accepts PSAs?	yes (contact Mary Harris)
Notes	Does not broadcast local news.

WWCA (1270 AM)

Address	107 W. 78th Pl., Merrillville IN 46410	
Telephone	219-736-7524	
E-mail	info@relevantradio.com	
Web Site	www.relevantradio.com	
Wattage	1,000	
Format	Catholic Talk Radio	
Network Affiliations	Relevant Radio	
On-air Hours	24/7	
Broadcast Company	Starboard Media Foundation	
General Mgr./News Dir.	Scott Wert	swert@relevantradio.com
Sales Manager	Bob Benes	bbenes@relevantradio.com
Accepts PSAs?	yes (contact Scott Wert)	

WYIN TV (Channel 56)

Address	8625 Indiana Pl., Merrillville IN 46410	
Telephone	219-756-5656	
Telephone (toll-free)	888-694-LAKE	
Fax	219-755-4312	
E-mail	info@lakeshorePTV.com	
E-mail (news)	news@lakeshorePTV.com	
Web Site	www.lakeshorePTV.com	
Network Affiliation	PBS	
On-air Hours	24/7	
Broadcast Company	Lakeshore Public Television	
President/CEO	Thomas Carroll	tcarroll@lakeshorePTV.com
News Director	Ryan Priest	rpriest@lakeshorePTV.com
Assistant News Director	Jodi Juhl	jjuhl@lakeshorePTV.com
News Assignment Mgr.	Lindsay Grome	lgrome@lakeshorePTV.com
Sports Director	Joe Arredondo	jarredondo@lakeshorePTV.com
Program Director	Julie Turkowski	jturkowski@lakeshorePTV.com
V. P. Sales	Renee Golas	rgolas@lakeshorePTV.com
Mrktg./Special Events Dir.	Megan Ciszewski	mciszewski@lakeshorePTV.com
Accepts PSAs?	yes-limited number (contact Kathleen Szot, kszot@lakeshorePTV.com)	
Notes	Non-commercial station.	

MICHIGAN CITY

LaPorte County

Beacher

Address	911 Franklin St., Michigan City IN 46360
Telephone	219-879-0088
Fax	219-879-8070
E-mail	beacher@thebeacher.com
E-mail (news)	sallym@thebeacher.com
Web Site	www.thebeacher.com
Publication Date	Weekly (Thursday)
Circulation	4,100 (free/delivered)
Publishing Company	Beacher Business Printers

News Dispatch

Address	121 W. Michigan Blvd., Michigan City IN 46360	
Telephone	219-874-7211	
Telephone (toll-free)	800-489-9292	
Fax (news)	219-872-8511	
Fax (advertising)	219-878-4487	
E-mail (news)	news@thenewsdispatch.com	
Web Site	www.thenewsdispatch.com	
Publication Date	Daily (Sunday-Saturday)	
Circulation	12,000 (paid)	
Publishing Company	Paxton Media Group	
Publisher	Patrick Kellar	pkellar@heraldargus.com
Executive Editor	Chris Schable	cschable@heraldargus.com
Managing Editor	Dave Hawk	dhawk@thenewsdispatch.com
Advertising Manager	Isis Cains	icains@thenewsdispatch.com
Sports Editor	Adam Parkhouse	aparkhouse@thenewsdispatch.com
Business Editor	Andrew Tallackson	atallackson@thenewsdispatch.com

WEFM (95.9 FM)

Address	1903 Springland Ave., Michigan City IN 46360
Telephone	219-879-8201
Fax	219-879-8202
E-mail	wefmr@yahoo.com
Wattage	4,000
Format	Adult Contemporary/Oldies (day); Jazz (night)
Network Affiliations	Brownfield, NBC, Network Indiana
On-air Hours	24/7
Broadcast Company	Michigan City FM Broadcasters
General Mgr./Sales Mgr.	Ron Miller
Program Director	Tod Allen
Sports Director	Paul Condry
Accepts PSAs?	yes (contact Tod Allen)

See **WFRN** (Elkhart).

WIMS (1420 AM)

Address	720 Franklin St., Michigan City IN 46360
Telephone	219-879-9810
Telephone (news)	219-879-9812
Fax	219-879-9813
E-mail	info@wimsradio.com
E-mail (news)	news@wimsradio.com
Web Site	www.wimsradio.com
Wattage	5,000
Format	News/Talk
On-air Hours	24/7
Broadcast Company	Gerard Media
General Manager	Ric Federighi ric@wimsradio.com
Program/Sports Director	Johnny Rush rush@wimsradio.com
Sales Manager	Jody Rogers jody@wimsradio.com
Accepts PSAs?	yes (contact Johnny Rush)

MIDDLEBURY

Elkhart County

See **Middlebury Independent** (LaGrange).

MIDDLETOWN

Henry County

Middletown News

Address	106 N. 5th St., Middletown IN 47356
Telephone/Fax	765-354-2221
E-mail	drew@themiddletownnews.com
Web Site	www.themiddletownnews.com
Publication Date	Weekly (Thursday)
Circulation	2,000 (paid)
Publisher	Drew Cooper drew@themiddletownnews.com
Editor	Joey Cooper joey@themiddletownnews.com

MILFORD

The Mail-Journal — 'the Paper'
Kosciusko County

Mailing Address	P.O. Box 188, Milford IN 46542
Street Address	206 S. Main St., Milford IN 46542
Telephone	574-658-4111
Telephone (toll-free)	800-733-4111
Fax	574-658-4701
E-mail (news)	jseely@the-papers.com
Publishing Company	The Papers Inc.
Publisher	Ron Baumgartner	rbaumgartner@the-papers.com
Editor	Jeri Seely	jseely@the-papers.com
Advertising Manager	Kip Schumm	kschumm@the-papers.com
Sports Editor	Mike Deak	mdeak@the-papers.com
Business Manager	Collette Knapp	cknapp@the-papers.com
Notes	Also publishes **Indiana Auto & RV, Indianapolis House & Home,** and **Michiana House & Home** (primarily all-advertising publications).

The Mail-Journal
Publication Date	Weekly (Wednesday)
Circulation	3,000 (paid)

'the Paper'
Publication Date	Weekly: Tuesday (Elkhart County edition)
	Weekly: Wednesday (Kosciusko County edition)
Circulation	33,000 (Elkhart County edition)
	22,700 (Kosciusko County edition)
Notes	Bureaus in Goshen, Syracuse and Warsaw.

See **Senior Life** (listed under Senior Citizens/Specialty Publications).

MISHAWAKA

Mishawaka Enterprise
Saint Joseph County

Mailing Address	P.O. Box 584, Mishawaka IN 46546
Street Address	419 W. 7th St., Mishawaka IN 46544
Telephone/Fax	574-255-4789
E-mail	mishawakanews@aol.com
Publication Date	Weekly (Thursday)
Circulation	2,000 (paid)
Publishing Company	ECOM Publishing
Editor	Bill Nich

South Bend Tribune—Bureau
Address	1301 E. Douglas Rd., Mishawaka IN 46545
Telephone	574-247-7754
Fax	574-289-0622
E-mail (news)	eball@sbtinfo.com
Web Site	www.southbendtribune.com
Mishawaka Editor	Ellen Ball	eball@sbtinfo.com
Notes	Main office in South Bend.

WAOR (95.3 FM) — WBYT (100.7 FM) — WRBR (103.9 FM) — WYPW (95.7 FM)

Address	237 W. Edison Rd., Mishawaka IN 46545
Telephone	574-258-5483
Telephone (news)	866-691-6397
Fax	574-258-0930
Fax (news)	574-257-2345
E-mail (news)	news@michiananewschannel.com
On-air Hours	24/7
News Director	Heather Richards news@michiananewschannel.com
Promotions Director	Kayla Ernsberger kernsberger@federatedmedia.com

WAOR

Web Site	www.waor.com
Wattage	5,500
Format	Classic Rock
Broadcast Company	Federated Media
General Manager	Brad Williams bwilliams@federatedmedia.com
Program Director	Mike Ragozino ragz@waor.com
Sales Manager	Stephanie Michel smichel@waor.com
Accepts PSAs?	yes (contact Deb Miles, deb@b100.com)

WBYT

Phone (toll-free)	888-817-2100
Web Site	www.b100.com
Wattage	15,000
Format	Country
Broadcast Company	Federated Media
General Manager	Brad Williams bwilliams@federatedmedia.com
Program Director	Mark Allen mark@b100.com
Sales Manager	Stephanie Michel smichel@waor.com
Accepts PSAs?	yes (contact Deb Miles, deb@b100.com)

WRBR

Web Site	www.wrbr.com
Wattage	3,000
Format	Active Rock
Broadcast Company	Secondwind Communications
General Manager	Kathy Uebler kuebler@federatedmedia.com
Program Director	Tommy Carroll tcarroll@wrbr.com
Sales Manager	Jeff Smith jsmith@wrbr.com
Notes	Does not accept PSAs.

WYPW

Phone (toll-free)	888-737-6244
Web Site	www.power957.com
Wattage	3,000
Format	Top 40 Rhythm
Broadcast Company	Talking Stick Communications
General Manager/ Sales Manager	Kathy Uebler kuebler@federatedmedia.com
Program Director	Mike Jackson mikejax@power957.com
Accepts PSAs?	yes (contact Mike Jackson)

See **WFRN** (Elkhart).

WHFB (99.9 FM) — WNSN (101.5 FM) — WSBT (960 AM) — WZOC (94.3 FM)

Address	1301 E. Douglas Rd., Mishawaka IN 46545
Telephone	574-233-3141
Telephone (news)	574-472-8257
Fax	574-239-4231
E-mail	pmorris@wsbt.com
E-mail (news)	montgom@wsbt.com
On-air Hours	24/7
General Manager	Sally Brown
News Director	Bob Montgomery	montgom@wsbt.com
Sports Director	Darin Pritchett	darin@wsbt.com
Public Affairs Director	Mary Simko	simko@wsbt.com
Promotions Director	Kimberly Crim	crim@wsbt.com
Traffic Director	Donna Snyder	donna@wsbt.com
Accepts PSAs?	yes (contact Mary Simko)

WHFB

Web Site	www.catcountry999.com	
Wattage	13,000	
Format	Country	
Network	Westwood One	
Broadcast Company	WHFB Broadcast Associates	
Program Director	Jim Roberts	roberts@wsbt.com
Notes	Serves Benton Harbor, MI.	

WNSN

Web Site	www.sunny1015.com	
Wattage	13,000	
Format	Adult Contemporary	
Network	Westwood One	
Broadcast Company	Schurz Communications	
Program Director	Jim Roberts	roberts@wsbt.com
Notes	Serves South Bend.	

WSBT

Web Site	www.wsbtradio.com	
Wattage	5,000	
Format	News/Talk	
Network	Fox News	
Broadcast Company	Schurz Communications	
Program Director	Bob Montgomery	montgom@wsbt.com
Notes	Serves South Bend.	

WZOC

Web Site	www.oldies943fm.com	
Wattage	11,500	
Format	Oldies	
Network	Dial Global	
Broadcast Company	Plymouth Broadcasting	
Program Director	Buddy King	buddy@buddyking.net
Notes	Serves South Bend & Plymouth. Has studio in Plymouth.	

WBND TV (Channel 57.1) — WCWW TV (Channel 25.1) — WMYS TV (Channel 23)

Address	3665 Park Pl. W. , STE 200, Mishawaka IN 46545
Telephone	574-243-4316
Fax	574-243-4326
Web Sites	www.abc57.com
	www.thecw25.com
	www.mymichianatv.com
Network Affiliations	ABC (WBND)
	The CW (WCWW)
	MyNetwork TV (WMYS)
On-air Hours	24/7
Broadcast Company	Weigel Broadcasting
General Manager	Jeff Guy	jguy@abc57.com
Local Sales Mgr. (WBND)	Frank Hawkins	fhawkins@abc57.com
Local Sales Mgr. (WCWW/WMYS)	Josh Bramer	jbramer@mymichianatv.com
Program Director	Kyle Hart	khart@wciu.com
Promotions Director	Jim Roche	jroche@wciu.com
Accepts PSAs?	yes (contact Carol Van Dusen, cvandusen@abc57.com)

WSBT TV (Channels 22.1/CBS; 22.2/SBT2; & 22.3/Stormtracker Channel)

Address	1301 E. Douglas Rd., Mishawaka IN 46545
Telephone	574-233-3141
Telephone (toll-free)	800-872-3141
Telephone (news)	574-232-6397
Fax	574-288-6630
Fax (news)	574-289-0622
E-mail (news)	wsbtnews@wsbt.com
Web Site	www.wsbt.com
Network Affiliation	CBS
On-air Hours	24/7
Broadcast Company	Schurz Communications
General Manager	John Mann	jmann@wsbt.com
News Director	Meg Sauer	msauer@wsbt.com
News Assignment Editor	Brian Sapp	bsapp@wsbt.com
Sports Director	Pete Byrne	pbyrne@wsbt.com
General Sales Manager	Beth Young	byoung@wsbt.com
Local Sales Manager	Gregg Richardson	greggr@wsbt.com
Program Director	Bob Johnson	johnson@wsbt.com
Public Affairs Director	Zane Torrence	zane@wsbt.com
Promotions Director	Scott Leiter	leiter@wsbt.com
Chief Engineer	C. Eugene Hale	hale@wsbt.com
Accepts PSAs?	yes (contact Cassie Hopkins, chopkins@wsbt.com)
Notes	Bureau in Berrien Springs, MI.

MITCHELL

Lawrence County

See **WMBL** (Anderson).

See **WPHZ** (Bedford).

MONON

News & Review
White County

Mailing Address	P.O. Box 98, Monon IN 47959
Street Address	106 W. 4th St., Monon IN 47959
Telephone	219-253-6234
Telephone (toll-free)	800-832-8791
Fax	219-253-2636
E-mail	newsandreview@urhere.net
Web Site	www.smalltownpapers.com
Publication Date	Weekly (Wednesday)
Circulation	1,100 (paid)
Publisher/Editor	Lori Breen

MONROEVILLE

Monroeville News
Allen County

Mailing Address	P.O. Box 429, Monroeville IN 46773
Street Address	115 E. South St., Monroeville IN 46773
Telephone	260-623-3316
Fax	260-623-3966
E-mail	loisternet@yahoo.com
Publication Date	Weekly (Wednesday)
Circulation	1,200 (paid)
Publishing Company	Decatur Publishing
Publisher	B. J. Riley
Editor	Lois Ternet

MONTICELLO

Herald Journal
White County

Mailing Address	P.O. Box 409, Monticello IN 47960	
Street Address	114 S. Main St., Monticello IN 47960	
Telephone	574-583-5121	
Telephone (toll-free)	800-541-7906	
Fax	574-583-4241	
E-mail (news)	twright@thehj.com	
Web Site	www.thehj.com	
Publication Date	Daily (Monday-Saturday)	
Circulation	4,600 (paid)	
Publishing Company	Kankakee Valley Publishing Co.	
Publisher	Don Hurd	dongo75@aol.com
Editor	Trent Wright	twright@thehj.com
Advertising Manager	Karen Franscoviak	kfranscoviak@thehj.com
Sports Editor	Chad Husted	chusted@thehj.com
TMC/Shopper	The Reminder (weekly/serving Carroll and White counties)	

WMRS (107.7 FM)

Address	132 N. Main St., Monticello IN 47960
Telephone/Fax	574-583-8933
E-mail	kevinp@wmrsradio.com
Web Site	www.wmrsradio.com
Wattage	6,000
Format	Adult Contemporary
Network Affiliations	Brownfield, Colts, Jones, Pacers, USA Network News, White Sox
On-air Hours	24/7
Broadcast Company	Monticello Community Radio
General Manager	Kevin Page kevinp@wmrsradio.com
News/Promotions Director	Jaime Page jaime@wmrsradio.com
Program/Sports Director	Brandi Page brandi@wmrsradio.com
Sales Manager	Laura Page laura@wmrsradio.com
Public Affairs Director	Shirley Grigsby
Accepts PSAs?	yes (contact Shirley Grigsby)

MONTPELIER
Blackford County

See **WJCO** (Valparaiso).

MOORESVILLE
Morgan County

Mooresville/Decatur Times

Address	23 E. Main St., Mooresville IN 46158
Telephone	317-831-0280
Fax	317-831-7068
Fax (advertising)	317-831-7070
E-mail (news)	ahillenb@md-times.com
Web Site	www.md-times.com
Publication Date	Semi-weekly (Wednesday & Saturday)
Circulation	3,500 (paid)
Publishing Company	Schurz Communications
Publisher	E. Mayer Maloney Jr. mmaloney@md-times.com
Managing Editor	Brian Culp bculp@reporter-times.com
Advertising Manager	Cory Bollinger cbollinger@reportert.com
Sports Editor	Steve Page spage@md-times.com
News Editor	Amy Hillenburg ahillenb@md-times.com

MOROCCO
Newton County

See **Morocco** Courier (Kentland).

MORRISTOWN
Shelby County

See **WJCF** (Greenfield).

MOUNT COMFORT
Hancock County

See **WJCF** (Greenfield).

MOUNT VERNON

Posey County

Mount Vernon Democrat

Mailing Address	P.O. Box 767, Mount Vernon IN 47620
Street Address	231 Main St., STE A, Mount Vernon IN 47620
Telephone	812-838-4811
Fax	812-838-3696
Web Site	www.mvdemocrat.com
Publication Date	Weekly (Wednesday)
Circulation	4,000 (paid)
Publishing Company	Landmark Community Newspapers
Publisher	Kevin Lashbrook publisher@perrycountynews.com
Editor	Angela Geralds editor@mvdemocrat.com
Advertising Representative	Kim Tanner advertising@mvdemocrat.com
Sports Writer	Cory Woolsey sports@mvdemocrat.com
TMC/Shopper	Posey Advantage (weekly)

WRCY (1590 AM) & WYFX (106.7 FM)—Studio

Address	7109 Upton Rd., Mount Vernon IN 47620
Telephone	812-838-4484
Fax	812-838-6434
Operations Manager	Sean Dulaney seandulaney@originalcompany.com
Notes	Main office in Vincennes.

MUNCIE

Delaware County

See **Ball State Daily News** (listed under College Campus/Specialty Publications).

Muncie Times

Address	1304 N. Dr. Martin Luther King Jr. Blvd., Muncie IN 47303
Telephone	765-741-0037
Fax	765-741-0040
E-mail	themuncietimes@comcast.net
Publication Date	Semi-monthly
Circulation	5,000 (free)
Publisher	Beatrice Moten-Foster
Editor	T. Kumbula
Notes	Serves African-American community.

Star Press

Mailing Address	P.O. Box 2408, Muncie IN 47307
Street Address	345 S. High St., Muncie IN 47305
Telephone	765-213-5700
Telephone (toll-free)	800-783-3932
Telephone (news)	765-213-5754
Fax	765-213-5703
Fax (news)	765-213-5858
Fax (advertising)	765-213-5937
E-mail (news)	news@muncie.gannett.com
Web Site	www.thestarpress.com
Publication Date	Daily (Sunday-Saturday)
Circulation	30,400-paid (daily); 32,000-paid (Sunday)
Publishing Company	Gannett Co. Inc.
General Mgr./Exec. Editor	Gene Williams gwilliams@muncie.gannett.com
Advertising Manager	Mary Young myoung@muncie.gannett.com
Sports Editor	Greg Fallon gfallon@muncie.gannett.com

See **WBCL** (Fort Wayne).

WBSB (89.5 FM) — **WBSH** (91.1 FM) — **WBSJ** (91.7 FM) — **WBST** (92.1 FM) **WBSW** (90.5 FM)

Address	LB 128, Ball State University, Muncie IN 47306
Telephone	765-285-5888
Telephone (toll-free)	800-646-1812
Telephone (news)	765-285-8999
Fax	765-285-8937
Fax (news)	765-285-6397
E-mail	ipr@bsu.edu
E-mail (news)	bmbeaver@bsu.edu
Web Site	www.bsu.edu/ipr
Wattage	400 (WBSB)
	300 (WBSH)
	2,000 (WBSJ)
	3,000 (WBST)
	1,000 (WBSW)
Format	American Public Media, NPR, PRI
Network Affiliations	NPR, PRI
On-air Hours	24/7
Owner	Ball State University
General Manager	Marcus Jackman mjackman@bsu.edu
News Director	Terry Heifetz tjheifetz@bsu.edu
Music Director	Steven Turpin sturpin@bsu.edu
Outreach Coordinator	Carol Trimmer ctrimmer@bsu.edu
Accepts PSAs?	yes (contact Carol Trimmer)
Notes	Non-commercial stations. WBSB (Anderson), WBSH (New Castle), WBSJ (Portland), and WBSW (Marion) are a simulcast of WBST.

WCRD (91.3 FM)

Address	LB 200, Ball State University, Muncie IN 47306
Telephone	765-285-1467
Fax	765-285-9278
E-mail	wcrd@bsu.edu
Web Site	www.wcrd.net
Wattage	310
Format	Variety
On-air Hours	24/7
Owner	Ball State University
Faculty Advisor	Barry Umansky bdumansky@bsu.edu
Accepts PSAs?	yes
Notes	Non-commercial station. All positions filled by students.

WERK (104.9 FM) — **WHBU** (1240 AM) — **WLBC** (104.1 FM) — **WMQX** (96.7 FM)
WMXQ (93.5 FM) — **WXFN** (1340 AM)

Address	800 E. 29th St., Muncie IN 47302
Telephone	765-288-4403
Telephone (news)	765-289-6397
Telephone (Daleville studio)	765-378-2080
Fax	765-288-0429
E-mail (news)	crystal.cole@bybradio.com
On-air Hours	24/7
Broadcast Company	Backyard Broadcasting
General Manager	Bruce Law bruce.law@bybradio.com
V.P./Operations Manager	Steve Lindell steve.lindell@bybradio.com
News Director	Crystal Cole crystal.cole@bybradio.com

WERK

Phone (toll-free)	866-516-1278
E-mail	daleville.studio@bybradio.com
Web Site	www.werkradio.com
Wattage	6,000
Format	Oldies
Program Director	Leland Franklin leland.franklin@bybradio.com
Sales Manager	Preston Corey preston.corey@bybradio.com
Accepts PSAs?	yes (contact Leland Franklin)

WHBU

Phone (toll-free)	800-578-1240
E-mail	daleville.studio@bybradio.com
Web Site	www.1240whbu.com
Wattage	1,000
Format	News/Talk
Network	CBS News
Program Director	Leland Franklin leland.franklin@bybradio.com
Sales Manager	Preston Corey preston.corey@bybradio.com
Accepts PSAs?	yes (contact Steve Lindell)
Notes	Serves Anderson.

WLBC

Phone (toll-free)	866-317-9522
E-mail	wlbc.studio@bybradio.com
Web Site	www.wlbc.com
Wattage	50,000
Format	Adult Contemporary
Program Director	Steve Lindell steve.lindell@bybradio.com
Sales Manager	Amy Dillon amy.dillon@bybradio.com
Accepts PSAs?	yes (contact Steve Lindell)

WMQX & WMXQ

E-mail	daleville.studio@bybradio.com
Web Site	www.maxrocks.net
Wattage	3,000 (WMQX)
	6,000 (WMXQ)
Format	Classic Rock
Program Director	Jay Garrison jay.garrison@bybradio.com
Sales Manager	Preston Corey preston.corey@bybradio.com
Accepts PSAs?	yes (contact Jay Garrison)
Notes	WMQX & WMXQ are simulcast. Stations serve Anderson.

WXFN

E-mail	wlbc.studio@bybradio.com
Wattage	1,000
Format	Sports
Network	ESPN
On-air Hours	24/7
Program Director	Steve Lindell	steve.lindell@bybradio.com
Sales Manager	Amy Dillon	amy.dillon@bybradio.com
Accepts PSAs?	yes (contact Steve Lindell)
Notes	Serves Anderson.

See **WJCF** (Greenfield).

See **WKMV/K-Love** (listed under National Radio Stations).

See **WMDH FM** (New Castle).

See **WRFM AM** (Greenfield).

WIPB TV (Channels 49.1/PBS & 49.2/PBS Create)

Address	Edmund F. Ball Bldg., Ball State U., Muncie IN 47306
Telephone	765-285-1249
Telephone (toll-free)	800-252-9472
Fax	765-285-5548
E-mail	wipb@bsu.edu
Web Site	www.bsu.edu/wipb
Network Affiliation	PBS
On-air Hours	24/7
Owner	Ball State University
General Manager	Alice Cheney
Program Director	Sue Bunner
Accepts PSAs?	yes (contact Sue Bunner)
Notes	Non-commercial station. Does not broadcast local news.

WMUN TV (Channel 26)

Mailing Address	P.O. Box 2357, Muncie IN 47307
Telephone	765-759-7975
E-mail	actstv@att.net
Network Affiliation	TBN
On-air Hours	24/7
Owner	Full Gospel Business Men's Fellowship
General Manager	Donald Badgley
Accepts PSAs?	no
Notes	Does not broadcast local news.

MUNSTER

The Times of Northwest Indiana
Lake County

Address	601 W. 45th Ave., Munster IN 46321
Telephone	219-933-3200
Telephone (toll-free)	800-589-3221
Telephone (news)	219-933-3223
Fax (news)	219-933-3249
Fax (advertising)	219-933-3332
Web Site	www.nwi.com
Publication Date	Daily (Sunday-Saturday)
Circulation	89,000-paid (daily); 96,000-paid (Sunday)
Publishing Company	Lee Enterprises
Publisher	Bill Masterson Jr.	bill.masterson@nwitimes.com
Executive Editor	William Nangle	nangle@nwitimes.com
Managing Editor	Paul Mullaney	mullaney@nwitimes.com
Advertising Manager	Lisa Daugherty	ldaugherty@nwitimes.com
Sports Editor	Justin Breen	jbreen@nwitimes.com
Editorial Page Editor	Doug Ross	doug.ross@nwitimes.com
TMC/Shopper	TMC (weekly)	
Notes	Bureaus in Crown Point, Indianapolis, Portage & Valparaiso. Publishes 4 editions.

NAPPANEE
Elkhart County

See **Advance News** (Bremen).

NASHVILLE
Brown County

Brown County Democrat

Mailing Address	P.O. Box 277, Nashville IN 47448
Street Address	147 E. Main St., Nashville IN 47448
Telephone	812-988-2221
Fax	812-988-6502
E-mail	newsroom@bcdemocrat.com
Web Site	www.bcdemocrat.com
Publication Date	Weekly (Wednesday)
Circulation	4,800 (paid)
Publishing Company	Home News Enterprises
General Manager	Steve Marshall	smarshall@bcdemocrat.com
Editor	Linda Margison	lmargison@bcdemocrat.com
Advertising Manager	Keith Fleener	kfleener@bcdemocrat.com
TMC/Shopper	The Marketplace (weekly)	

NEW ALBANY

Floyd County

Tribune

Mailing Address	P.O. Box 997, New Albany IN 47151
Street Address	303 Scribner Dr., New Albany IN 47150
Telephone	812-944-6481
Fax (news)	812-949-6585
E-mail (news)	newsroom@newsandtribune.com
Web Site	www.newsandtribune.com
Publication Date	Daily (Tuesday-Sunday)
Circulation	7,000 (paid)
Publishing Company	Community Newspaper Holdings Inc.
Publisher/Executive Editor	Steve Kozarovich steve.kozarovich@newsandtribune.com
Managing Editor	Shea Van Hoy shea.vanhoy@newsandtribune.com
Advertising Manager	Mary Tuttle mary.tuttle@newsandtribune.com
Sports Editor	Mike Hutsell sports@newsandtribune.com

See **Air 1** (listed under National Radio Stations).

WNAS (88.1 FM)

Address	New Albany High School, 1020 Vincennes St., New Albany IN 47150
Telephone	812-981-7621
Fax	812-949-6926
E-mail	lkelly@wnas.org
Web Site	www.wnas.org
Wattage	2,850
Format	Eclectic
On-air Hours	24/7
Owner	New Albany/Floyd County Consolidated School Corp.
General Manager	Lee Kelly lkelly@wnas.org
Accepts PSAs?	yes (contact Lee Kelly)
Notes	Non-commercial station.

NEW CASTLE

Henry County

Courier-Times

Mailing Address	P.O. Box 369, New Castle IN 47362
Street Address	201 S. 14th St., New Castle IN 47362
Telephone	765-529-1111
Telephone (toll-free)	800-489-2472
Fax	765-529-1731
E-mail (news)	editor@thecouriertimes.com
Web Site	www.thecouriertimes.com
Publication Date	Daily (Monday-Saturday)
Circulation	9,500 (paid)
Publishing Company	Paxton Media Group
Publisher	Tina West twest@thecouriertimes.com
Managing Editor	Randy Rendfeld editor@thecouriertimes.com
Advertising Manager	Scott Hart shart@thecouriertimes.com
Sports Editor	Jeremy Hines jhines@thecouriertimes.com
TMC/Shopper	CT Extra (weekly)

See **WBSH** (Muncie).

See **WJCF** (Greenfield).

WMDH (1550 AM) — WMDH (102.5 FM)

Mailing Address	P.O. Box 690, New Castle IN 47362
Street Address	1134 W. State Rd. 38, New Castle IN 47362
Telephone	765-529-2600
Fax	765-529-1688
Web Site	www.wmdh.com
On-air Hours	24/7
Broadcast Company	Citadel Broadcasting Co.
General Manager	Todd Merickel todd.merickel@citcomm.com
Program Director	Shane Goad shane.goad@citcomm.com
Sales Manager	Pam Price pam.price@citcomm.com
Public Affairs/Promotions	Cara Denis cara.denis@citcomm.com
Accepts PSAs?	yes (contact Cara Denis)

WMDH AM
Wattage	250
Format	Real Country
Network	ABC

WMDH FM
Wattage	50,000
Format	Country
Networks	Learfield, Westwood One
Notes	Serves New Castle, Muncie & Anderson.

NEW HARMONY
Posey County

Posey County News
Mailing Address	P.O. Box 397, New Harmony IN 47631
Street Address	641 3rd St., New Harmony IN 47631
Telephone	812-682-3950
Fax	812-682-3944
Web Site	www.poseycountynews.com
Publication Date	Weekly (Tuesday)
Circulation	4,000 (paid)
Owner/Publisher/Editor	Dave Pearce dpearce263@aol.com
Sports Editor	Steve Joos sports801@sbcglobal.net

NEW HAVEN
Allen County

See **East Allen County Times** (Fort Wayne).

NEW PALESTINE

Hancock County

New Palestine Press

Mailing Address	P.O. Box 407, New Palestine IN 46163
Street Address	25 W. Mill St., New Palestine IN 46163
Telephone	317-861-4242
Fax	317-861-4201
E-mail	nppress407@aol.com
Publication Date	Weekly (Thursday)
Circulation	2,500 (paid)
Owner/Publisher	Rebecca Gaines
Editor	Vickie Williams
Advertising Manager	Gary Hartman

See **New Palestine Reporter** (Greenfield).

NEWBURGH

Warrick County

See **Newburgh Register** (Boonville).

NOBLESVILLE

Hamilton County

See **Current in Noblesville** (Carmel).

The Times

Address	641 Westfield Rd., Noblesville IN 46060	
Telephone	317-770-7777	
Fax	317-770-9376	
E-mail	news@thetimes24-7.com	
Web Site	www.thetimes24-7.com	
Publication Date:	Daily (Monday & Wednesday-Saturday)	
Circulation	8,000	
Publishing Company	The Paper of Montgomery County	
Publisher/Editor	Tim Timmons	ttimmons@thetimes24-7.com
Advertising Manager	David Thornberry	dthornberry@thetimes24-7.com

HomeTown Sports Indiana (Cable-only Channel 81)
HomeTown Television (Cable-only Channel 19)

Mailing Address	P.O. Box 1386, Noblesville IN 46061
Street Address	92 S. 9th St., Noblesville IN 46060
Telephone/Fax	317-770-4670
Network Affiliation	independent
Broadcast Company	HomeTown Television Corp.
General Manager	Roy Johnson roy@hometowntelevision.com
Program Director	Allan Hughes
Notes	Does not broadcast local news.

HomeTown Sports Indiana

E-mail	info@HometownSportsIndiana.com
Web Site	www.HometownSportsIndiana.com
Sports/Sales Mgr.	Greg Rakestraw greg@hometownsportsindiana.com
Accepts PSAs?	no

HomeTown Television

E-mail	info@hometowntelevision.com
Web Site	www.HometownTelevision.com
Sales Manager	Roy Johnson roy@HometownTelevision.net
Accepts PSAs?	yes (contact Roy Johnson)

WHMB TV (Channels 40.1/main programming & 40.2/World Harvest Television)

Address	10511 Greenfield Ave., Noblesville IN 46060
Telephone	317-773-5050
Fax	317-776-4051
Web Site	www.whmbtv.com
Network Affiliation	independent
On-air Hours	24/7
Broadcast Company	LeSEA Broadcasting of Indianapolis, Inc.
General Mgr./Sales Mgr.	Keith Passon kpasson@lesea.com
Sports Director	Dennis Kasey dkasey@lesea.com
Public Affairs Director	Karen Robinson krobinson@lesea.com
Accepts PSAs?	yes (contact Karen Robinson)
Notes	Does not broadcast local news.

NORTH MANCHESTER

Wabash County

News-Journal

Mailing Address	P.O. Box 368, North Manchester IN 46962
Street Address	1306 State Road 114 W., North Manchester IN 46962
Telephone	260-982-6383
Fax	260-982-8233
E-mail	info@nmpaper.com
E-mail (news)	news@nmpaper.com
Web Site	www.nmpaper.com
Publication Date	Weekly (Wednesday)
Circulation	1,800 (paid)
Publisher	Mike Rees mrees@thepaperofwabash.com
Managing Editor	Eric Christiansen news@nmpaper.com
Advertising Manager	Carrie Vineyard ads@nmpaper.com

WBKE (89.5 FM)

Mailing Address	Manchester College, 604 College Ave., MC Box 19, North Manchester IN 46962
Street Address	Manchester College, 604 College Ave., North Manchester IN 46962
Telephone	260-982-5272
Fax	260-982-5043
E-mail	ddaggett@manchester.edu
Web Site	http://wbke.manchester.edu
Wattage	820
Format	College Radio
Network Affiliations	NPR
On-air Hours	24/7
Owner	Manchester College
Station Advisor	Dan Daggett ddaggett@manchester.edu
Accepts PSAs?	yes (contact Dan Daggett)
Notes	Non-commercial station.

NORTH VERNON

Jennings County

North Vernon Plain Dealer — North Vernon Sun

Mailing Address	P.O. Box 988, North Vernon IN 47265
Street Address	528 E. O & M Ave., North Vernon IN 47265
Telephone	812-346-3973
Fax	812-346-8368
E-mail	pds@northvernon.com
Web Site	www.plaindealer-sun.com
Publication Date	Weekly: Thursday (North Vernon Plain Dealer)
	Weekly: Tuesday (North Vernon Sun)
Circulation	8,000-paid (North Vernon Plain Dealer)
	6,200-paid (North Vernon Sun)
Publisher	Barbara King bking@northvernon.com
Editor	Bryce Mayer bmayer@northvernon.com
Advertising Manager	Joshua Taylor jtaylor@northvernon.com
Sports Editor	Sharon Hamilton shamilton@northvernon.com

WJCP (1460 AM)

Address	2470 N. State Hwy. 7, North Vernon IN 47265
Telephone	812-346-9527
Fax	812-346-9722
E-mail	wjcp927@yahoo.com
Wattage	1,000
Format	Oldies
Network Affiliations	Citadel
On-air Hours	24/7
Broadcast Company	Columbus Radio Inc.
General Manager	Joe Ammerman joe_ammerman@yahoo.com
News Director	Albert Stormes wjcp927@yahoo.com
Accepts PSAs?	yes (contact Joe Ammerman)

NOTRE DAME

Saint Joseph County

See **Blue & Gold Illustrated** (listed under Sports/Specialty Publications).

See **Irish Sports Report** (listed under Sports/Specialty Publications).

See **Observer** (listed under College Campus/Specialty Publications).

WSND (88.9 FM)

Address	315 LaFortune Student Center, Notre Dame IN 46556
Telephone	574-631-4069
Telephone (studio)	574-631-7342
Fax	574-631-3653
E-mail	wsnd@nd.edu
Web Site	www.nd.edu/~wsnd
Wattage	3,430
Format	Classical/Variety
On-air Hours	7:00 a.m. - 2:00 a.m. (Monday-Friday)
	9:00 a.m. - 2:00 a.m. (Saturday)
	8:00 a.m.-2:00 a.m. (Sunday)
Broadcast Company	Voice of the Fighting Irish, Inc.
General Manager	Laurie McFadden lmcfadde@nd.edu
Program Director	Ed Jaroszewski ejarosze@nd.edu
Accepts PSAs?	yes—only written (contact Ed Jaroszewski)
Notes	Non-commercial station. News from UPI Newswire

OAKLAND CITY

Gibson County

See **Oakland City Journal** (Princeton).

ODON

Daviess County

Odon Journal

Mailing Address	P.O. Box 307, Odon IN 47562
Street Address	102 W. Main St., Odon IN 47562
Telephone	812-636-7350
Fax	812-636-7359
E-mail	journal@rtccom.net
Publication Date	Weekly (Wednesday)
Circulation	3,000 (paid)
Publisher/Editor	John Myers journal1@rtccom.net
General Manager	Sue Myers

ORLAND

Steuben County

See **WCKZ** (Fort Wayne).

ORLEANS

Orange County

Progress Examiner
Mailing Address	P.O. Box 225, Orleans IN 47452
Street Address	233 S. 2nd St., Orleans IN 47452
Telephone/Fax	812-865-3242
E-mail	penews@blueriver.net
Publication Date	Weekly (Wednesday)
Circulation	2,200 (paid)
Publisher/Editor	John F. Noblitt
Associate Editor	Nancy Wright

OSSIAN

Wells County

Ossian Journal — Sunriser News
Mailing Address	P.O. Box 365, Ossian IN 46777
Street Address	1002 Dehner Dr., Ossian IN 46777
Telephone	260-622-4108
Fax	260-622-6439
Fax (news & advertising)	260-824-0700
E-mail	email@news-banner.com
Web Site	www.news-banner.com
Publication Date	Weekly: Thursday (Ossian Journal)
	Weekly: Tuesday (Sunriser News)
Circulation	700-paid (Ossian Journal)
	7,800-free/delivered (Sunriser News)
Publishing Company	News-Banner Publications
Publisher	Mark Miller miller@news-banner.com
Advertising Manager	Jean Bordner jeanb@news-banner.com
Advertising Representative	Chuck King king57@news-banner.com
Notes	Main office in Bluffton.

OXFORD

Benton County

WIBN (98.1 FM)
Mailing Address	P.O. Box 25, Oxford IN 47971
Street Address	130 E. McConnell, Oxford IN 47971
Telephone	765-385-2373
Fax	765-385-2374
Web Site	www.981wibn.com
Wattage	25,000
Format	Oldies
On-air Hours	24/7
Broadcast Company	Brothers Broadcasting, Inc.
General Manager	John Balvich
Operations Manager	Dan McKay danmckayporat@gmail.com
Accepts PSAs?	yes

PALMYRA

See **WYGS** (Columbus).

Harrison County

PAOLI

Orange County

Paoli News — Paoli Republican

Mailing Address	P.O. Box 190, Paoli IN 47454
Street Address	131 N.W. Court St., Paoli IN 47454
Telephone	812-723-2572
Telephone (toll-free)	888-884-5553
Fax	812-723-2592
E-mail	ocpinc@ocpnews.com
Web Site	www.paolinewsrepublican.com
Publication Date	Weekly: Thursday (Paoli News)
	Weekly: Tuesday (Paoli Republican)
Circulation	3,100-paid (Paoli News)
	3,200-paid (Paoli Republican)
Publishing Company	Orange County Publishing
Publisher/Editor	Arthur Hampton
Advertising Manager	Peggy Manship
TMC/Shopper	Orange Countian (weekly)

WKLO (96.9 FM) — WSEZ (1560 AM) — WUME (95.3 FM)

Mailing Address	P.O. Box 26, Paoli IN 47454
Street Address	192 S. Court St., Paoli IN 47454
Telephone	812-723-4484
Fax	812-723-4966
Broadcast Company	Diamond Shores Broadcasting
General Mgr./Sales Mgr.	Jerry Wall
News/Sports Director	Dave Dedrick
Program/Promotions Dir.	Jason Archer
Accepts PSAs?	yes (contact Matthew Montgomery)

WKLO

E-mail	wklo@blueriver.net
Web Site	www.realcountryonline.com
Wattage	6,000
Format	Real Country
Network	ABC
On-air Hours	24/7
Notes	Serves Hardinsburg.

WSEZ

Wattage	250
Format	Oldies
On-air Hours	7:00 a.m. - 5:00 p.m.
Notes	Does not broadcast local news.

WUME

E-mail	wume@blueriver.net
Web Site	www.wumemix95.com
Wattage	3,000
Format	Adult Contemporary
Network	ABC
On-air Hours	24/7

PEKIN

Banner-Gazette — Giveaway — Leader
Scott County Journal/Chronicle — Washington County Edition

Mailing Address	P.O. Box 38, Pekin IN 47165
Street Address	490 E. S.R. 60, Pekin IN 47165
Telephone	812-967-3176
Telephone (toll-free)	800-264-7336
Fax	812-967-3194
E-mail	paperman@gbpnews.com
Web Site	www.gbpnews.com
Publishing Company	Green Banner Publications
Publisher	Joe Green
Advertising Manager	April Falk
Notes	Bureaus in Charlestown, Salem & Scottsburg.

Banner-Gazette

Publication Date	Weekly (Wednesday)
Circulation	18,200 (free/delivered)
Editor	Joe Green

Giveaway

Publication Date	Weekly (Wednesday)
Circulation	19,000 (free/delivered)
Editor	Joe Green
Managing Editor	Marcus Amos
Notes	Serves Scott County, Henryville, Crothersville, eastern Jefferson County, and southern Jackson County.

Leader

Publication Date	Weekly (Wednesday)
Circulation	14,500 (free/delivered)
Editor	Joe Green
Managing Editor	Janna Ross
Notes	Serves Charlestown & Sellersburg.

Scott County Journal/Chronicle

Publication Date	Weekly (Saturday)
Circulation	5,000 (paid)
Editor	Marcus Amos
Notes	Serves Austin & Scottsburg.

Washington County Edition

Publication Date	Weekly (Wednesday)
Circulation	12,000 (free/delivered)
Editor	Joe Green
Notes	Serves Salem.

PENDLETON

See **Pendleton News** (Anderson).   Madison County

Times/Post

Mailing Address	P.O. Box 9, Pendleton IN 46064
Street Address	126 W. State St., Pendleton IN 46064
Telephone	765-778-2324
Fax	765-778-7152
E-mail	ptnews@ptlpnews.com
Publication Date	Weekly (Wednesday)
Circulation	3,500 (paid)
Publishing Company	Home News Enterprises
Publisher	Randall Shields	rdshields@greenfieldreporter.com
Editor	Jenny West	jennywest@ptlpnews.com
Advertising Representative	Christi Kincade	ckincade@ptlpnews.com
Notes	Serves Pendleton, Lapel and Markleville.

WEEM (91.7 FM)

Address	1 Arabian Dr., Pendleton IN 46064
Telephone	765-778-2161
Fax	765-778-0605
Web Site	www.917weem.org
Wattage	1,200
Format	Contemporary Hit Radio Mainstream/Top 40
Network Affiliations	Network Indiana
On-air Hours	24/7
Owner	South Madison Community School Corp.
General Manager	Jered Petrey	jpetrey@smadison.k12.in.us
Assistant General Manager	Chad Smith	csmith@smadison.k12.in.us
Accepts PSAs?	yes
Notes	Non-commercial station. Most positions filled by students.

PERU

Miami County

Peru Tribune

Mailing Address	P.O. Box 87, Peru IN 46970
Street Address	26 W. 3rd St., Peru IN 46970
Telephone	765-473-6641
Telephone (toll-free)	800-737-4488
Fax	765-472-4438
E-mail	ptads@perutribune.com
E-mail (news)	aturner@perutribune.com
Web Site	www.perutribune.com
Publication Date	Daily (Monday-Saturday)
Circulation	7,500 (paid)
Publishing Company	Paxton Media
Publisher	Randy Mitchell	rmitchell@paxtonmedia.com
Managing Editor	Aaron Turner	aturner@perutribune.com
Advertising Manager	Michelle Boswell	mboswell@perutribune.com
Sports Editor	Austan Kas	akas@perutribune.com
TMC/Shopper	Current Bargains (weekly)

WARU (1600 AM) — WARU (101.9 FM)

Mailing Address	P.O. Box 1010, Peru IN 46970
Street Address	1711 E. Wabash Rd., Peru IN 46970
Telephone	765-473-4448
Telephone (toll-free)	866-259-1019
Fax	765-473-4449
E-mail	waru@sbcglobal.net
Web Site	www.warufm.com
Wattage	1,000 (WARU AM)
	6,000 (WARU FM)
Format	Country
Network Affiliations	Fox, Westwood One
On-air Hours	24/7
Broadcast Company	Mid America Radio Group, Inc.
General Mgr./Sales Mgr.	Wade Weaver
News/Program Director	Andy McCord
Sports Director	Mark Ramsey
Public Affairs Director	Tammy Johnson
Accepts PSAs?	yes (contact Mark Ramsey)
Notes	WARU AM & FM are simulcast.

PETERSBURG

Pike County

Press-Dispatch

Mailing Address	P.O. Box 68, Petersburg IN 47567	
Street Address	820 E. Poplar St., Petersburg IN 47567	
Telephone	812-354-8500	
Fax	812-354-2014	
Publication Date	Weekly (Wednesday)	
Circulation	5,700 (paid)	
Publishing Company	Pike Publishing	
Publisher	Frank Heuring	
Editor	Andrew Heuring	editor@pressdispatch.net
Advertising Manager	John Heuring	jheuring@pressdispatch.net
Sports Editor	Mike Johansen	sports@pressdispatch.net

See **WBTO** (Vincennes).

PLAINFIELD

Hendricks County

Hendricks County ICON

Address	2680 E. Main St., STE 219, Plainfield IN 46168	
Telephone	317-837-5180	
Fax	317-837-4901	
E-mail	info@myicon.info	
Web Site	www.myicon.info	
Publication Date	Monthly	
Circulation	31,000 (free/mailed)	
Publishing Company	Times-Leader Publications, LLC	
Publisher	Rick Myers	rick@myicon.info
Notes	Mailed to all single-family homes in Avon, Brownsburg and Plainfield.	

PLYMOUTH

Bourbon News-Mirror — Culver Citizen — Pilot-News

Marshall County

Mailing Address	P.O. Box 220, Plymouth IN 46563
Street Address	214 N. Michigan St., Plymouth IN 46563
Telephone	574-936-3101
Telephone (toll-free)	800-933-0356
Fax (news)	574-936-3844
Fax (advertising)	574-936-7491
E-mail (news)	news@thepilotnews.com
Web Site	www.thepilotnews.com
Publishing Company	Horizon Publications
Publisher	Rick Kreps	rkreps@thepilotnews.com
General Manager	Jerry Bingle	jbingle@thepilotnews.com
Managing Editor	Maggie Nixon	mnixon@thepilotnews.com
Advertising Manager	Cindy Stockton	cstockton@thepilotnews.com
Sports Editor	Neil Costello	sports@thepilotnews.com
Notes	Also serves as the central office for the Advance News (Bremen), Bremen Enquirer (Bremen), and The Leader of Starke County (Knox).

Bourbon News-Mirror

Publication Date	Weekly (Thursday)
Circulation	900 (paid)
Editor	Angel Perkins
Notes	Serves Bourbon.

Culver Citizen

E-mail	citizen@culcom.net
Publication Date	Weekly (Thursday)
Circulation	1,200 (paid)
Editor	Jeff Kenney	citizen@culcom.net
Notes	Serves Culver.

Pilot-News

Publication Date	Daily (Monday-Saturday)
Circulation	5,500 (paid)
TMC/Shopper	The Shopper (weekly)

See **WIKV/K-Love** (listed under National Radio Stations).

WTCA (1050 AM)

Mailing Address	P.O. Box 519, Plymouth IN 46563
Street Address	112 W. Washington St., Plymouth IN 46563
Telephone	574-936-4096
Fax	574-936-6776
E-mail	wtca@am1050.com
E-mail (news)	news@am1050.com
Web Site	www.am1050.com
Wattage	250
Format	Classic Hits
On-air Hours	24/7
Broadcast Company	Community Service Broadcasters
General Mgr./News Dir.	Kathryn Bottorff
Program Director	Tony Ross
Sales Manager	Jim Bottorff
Accepts PSAs?	yes (contact Kathryn Bottorf)

See **WZOC** (Mishawaka).

PORTAGE

See **The Chronicle** (Valparaiso).   Porter County

The Times of Northwest Indiana—Bureau
Address   3410 Delta Dr., Portage IN 46368
Telephone   219-762-1397
Fax   219-762-8386
Notes   Main office in Munster.

WNDZ (750 AM)
Address   5625 N. Milwaukee Ave., Chicago IL 60646
Telephone   773-792-1121
Fax   773-792-2904
Web Site   www.accessradiochicago.com
Wattage   15,000
Format   Ethnic (brokered)
On-air Hours   sunrise - sunset
Broadcast Company   Newsweb Corp.
General Manager   Mark Pinski
News Director   Jorge Murillo
Sales Manager   Josh Fox
Public Affairs Director   John Poladian   outreachradio@gmail.com
Accepts PSAs?   yes
Notes   Serves Portage, IN & Chicago, IL

PORTLAND

Commercial Review   Jay County
Mailing Address   P.O. Box 1049, Portland IN 47371
Street Address   309 W. Main St., Portland IN 47371
Telephone   260-726-8141
Fax   260-726-8143
E-mail   cr.news@comcast.net
Web Site   www.thecr.com
Publication Date   Daily (Monday-Saturday)
Circulation   5,000 (paid)
Publishing Company   Graphic Printing Co.
Publisher/Editor   Jack Ronald
Managing Editor   Mike Snyder   cr.news@comcast.net
Advertising Manager Jeanne Lutz   cr.ads@comcast.net
Sports Editor   Ray Cooney   cr.sports@comcast.net
TMC/Shopper   The Circulator (weekly)

See **WBSJ** (Muncie).

WPGW (1440 AM) — WPGW (100.9 FM)

Address	1891 W. State Rd. 67, Portland IN 47371
Telephone	260-726-8780
Fax	260-726-4311
E-mail	wpgw@jayco.net
Wattage	540 (WPGW AM)
	4,600 (WPGW FM)
Format	Adult Contemporary (WPGW AM)
	Country (WPGW FM)
On-air Hours	6:00 a.m. - 10:00 p.m.
Broadcast Company	WPGW Inc.
General Mgr./Sales Mgr.	Rob Weaver
News Director	Jeff Overholser
Public Affairs Director	Laurette Horn
Accepts PSAs?	yes (contact Laurette Horn)

PRINCETON

Princeton Daily Clarion — Oakland City Journal

Gibson County

Mailing Address	P.O. Box 30, Princeton IN 47670
Street Address	100 N. Gibson St., Princeton IN 47670
Telephone	812-385-2525
Fax	812-386-6199
E-mail	news@pdclarion.com
Web Site	www.pdclarion.com
Publication Date	Weekly: Wednesday (Oakland City Journal)
	Daily: Monday-Friday (Princeton Daily Clarion)
Circulation	900-paid (Oakland City Journal)
	6,800-paid (Princeton Daily Clarion)
Publishing Company	Princeton Publishing Inc.
Publisher	Gary Blackburn — gblack@pdclarion.com
Editor	Andrea Howe — andrea@pdclarion.com
Advertising Manager	Tom Stephens — toms@pdclarion.com
Sports Editor	Pete Swanson — sports@pdclarion.com
TMC/Shopper	Gibson County Today (weekly)

WRAY (1250 AM) — WRAY (98.1 FM)

Mailing Address	P.O. Box 8, Princeton IN 47670
Street Address	1900 W. Broadway, Princeton IN 47670
Telephone	812-386-1250
Fax	812-386-6249
E-mail	wray@wrayradio.com
Web Site	www.wrayradio.com
Wattage	1,000 (WRAY AM)
	50,000 (WRAY FM)
Format	News/Talk (WRAY AM)
	Country (WRAY FM)
Network Affiliations	AP
On-air Hours	24/7
Broadcast Company	Princeton Broadcasting Co. Inc.
General Manager	Steve Lankford — steve@wrayradio.com
News Director	Cliff Ingram — cliff@wrayradio.com
Program Director	Dave Kunkel — dave@wrayradio.com
Sports Director	Jeff Lankford — jeff@wrayradio.com
Sales/Promotions Manager	Lynn Roach — lynn@wrayradio.com
Accepts PSAs?	yes (contact Lynn Roach)

WSJD (100.5 FM)

Address	328 Market St., Mt. Carmel IL 62863
Telephone	618-263-4300
Telephone (toll-free)	888-649-1005
Fax	618-263-3020
E-mail	wsjdgm@hotmail.com
Wattage	6,000
Format	Oldies
Network Affiliations	ABC, Chicago White Sox, IMS, Los Angeles Dodgers, Network Indiana, RFD Illinois, Sports USA Radio NFL Game of the Week
On-air Hours	24/7
General Mgr./Sales Mgr.	Kevin Madden wsjdgm@hotmail.com
Accepts PSAs?	yes
Notes	Serves Princeton and Evansville.

REMINGTON

Jasper County

See **Remington Press** (Rensselaer).

RENSSELAER

Jasper County

Remington Press — Rensselaer Republican

Address	117 N. Van Rensselaer St., Rensselaer IN 47978
Telephone	219-866-5111
Fax	219-866-3775
Publishing Company	Community Media Group
Publisher	Don Hurd dhurd@intranix.com
Editor	Clayton Doty editor@rensselaerrepublican.com
Sports Editor	Harley Tomlinson harley@rensselaerrepublican.com

Remington Press

Publication Date	Weekly (Wednesday)
Circulation	400 (paid)
TMC/Shopper	The Guide (weekly)
Notes	Serves Remington.

Rensselaer Republican

Web Site	www.myrepublican.info
Publication Date	Daily (Monday-Saturday)
Circulation	2,500 (paid)
TMC/Shopper	Shoppers News (weekly)

See **WIVR** (Kentland).

WLQI (97.7 FM) — WRIN (1560 AM)

Mailing Address	P.O. Box D, Rensselaer IN 47978
Street Address	560 W. Amsler Rd., Rensselaer IN 47978
Fax	219-866-5106
Broadcast Company	Brothers Broadcasting, Inc.
General Mgr./Sales Mgr.	John Balvich
News/Sports Director	Bob Kurtz
Program Director	Bob Burt
Accepts PSAs?	yes (contact Bob Burt)

WLQI

Telephone	219-866-4104
E-mail	wlqi@ffni.com
Web Site	www.977wlqi.com
Wattage	3,300
Format	Classic Hits
On-air Hours	24/7

WRIN

Telephone	219-866-5105
E-mail	wrin@ffni.com
Web Site	www.1560wrin.com
Wattage	1,000
Format	Easy Listening
On-air Hours	6:00 a.m. - sunset

WPUM (93.3 FM)

Mailing Address	P.O. Box 651, Rensselaer IN 47978
Street Address	U.S. Hwy. 231, St. Joseph's College, Rensselaer IN 47978
Telephone	219-866-6211
E-mail	wpum@saintjoe.edu
Web Site	www.saintjoe.edu/~wpum
Wattage	90
Format	Mix/Variety
On-air Hours	24/7
Owner	Saint Joseph's College
General Manager	Sally Berger sallyn@saintjoe.edu

RICHMOND

See **Earlham Word** (listed under College Campus/Specialty Publications).

Wayne County

Palladium-Item

Address	1175 N. "A" St., Richmond IN 47374
Telephone	765-962-1575
Telephone (toll-free)	800-686-1330
Telephone (news)	765-973-4474
Fax	765-973-4420
Fax (news)	765-973-4570
Fax (advertising)	765-973-4440
E-mail	palitem@pal-item.com
Web Site	www.pal-item.com
Publication Date	Daily (Sunday-Saturday)
Circulation	19,600-paid (daily); 24,000-paid (Sunday)
Publishing Company	Gannett Co. Inc.

General Manager/ Executive Editor	Mickey Johnson	mjohnso@pal-item.com
Managing Editor	Brian Guth	bguth@pal-item.com
Advertising Manager	Paige O'Neal	poneal@pal-item.com
Sports Editor	Josh Chapin	jchapin@pal-item.com

See **K-Love** (listed under National Radio Stations).

WECI (91.5 FM)

Address	Earlham College, 801 National Rd. W., Richmond IN 47374
Telephone	765-983-1246
Fax	765-983-1641
E-mail (news)	news@weciradio.org
E-mail (promotions)	publicity@weciradio.org
Web Site	www.weciradio.org
Wattage	400
Format	Eclectic
Network Affiliations	Pacifica
On-air Hours	24/7
Owner	Earlham College
Faculty Advisor	Avis Stewart
Accepts PSAs?	yes
Notes	Non-commercial station. Most positions filled by students.

WFMG (101.3 FM) — **WKBV** (1490 AM) — **WZZY** (98.3 FM)

Mailing Address	P.O. Box 1646, Richmond IN 47375
Street Address	2301 W. Main St., Richmond IN 47374
Telephone	765-962-6533
Telephone (toll-free)	877-983-9833
Fax	765-966-1499
E-mail (news)	bobphillips@G1013.com
On-air Hours	24/7
Broadcast Company	Whitewater Broadcasting
General Mgr./Sales Mgr.	Steve Frey steve@G1013.com
News Director	Bob Phillips bobphillips@G1013.com
Program Director	Rick Duncan rickduncan@G1013.com
Public Affairs Director	Jessica Leigh jessica@G1013.com
Promotions Director	Dave Snow davesnow@g1013.com
Accepts PSAs?	yes (contact Jessica Leigh)

WFMG

E-mail	info@G1013.com
Web Site	www.G1013.com
Wattage	20,500
Format	Hot Adult Contemporary

WKBV

Wattage	1,000
Format	News Talk
Networks	ABC, Brownfield Ag, ESPN, Network Indiana

WZZY

Web Site	www.todaysmusicmix.com
Wattage	3,000
Format	Adult Contemporary
Networks	Brownfield Ag, Pacers, Purdue
Notes	Serves Winchester.

WHON (930 AM) — **WQLK** (96.1 FM)

Mailing Address	P.O. Box 1647, Richmond IN 47375
Street Address	2626 Tingler Rd., Richmond IN 47374
Telephone	765-962-1595
Fax	765-966-4824
E-mail (news)	jeffL@kicks96.com
On-air Hours	24/7
Broadcast Company	Brewer Broadcasting
General Manager	Dave Strycker	daves@kicks96.com
News Director	Jeff Lane	jeffL@kicks96.com
Sales Manager	Paula Kay King	paulak@kicks96.com
Sports/Public Affairs Dir.	Troy Derengowski	troyd@930whon.com
Promotions Director	Abby Clapp	abbyc@kicks96.com
Accepts PSAs?	yes (contact Abby Clapp, abbyc@kicks96.com)

WHON

Phone (toll-free)	888-310-9466
Web Site	www.930whon.com
Wattage	500
Format	News/Talk/Sports
Networks	ABC News, Bloomberg, Fox Sports, Wall Street Journal
Program Director	Troy Derengowski	troyd@930whon.com

WQLK

Phone (toll-free)	877-542-5796
Web Site	www.kicks96.com
Wattage	50,000
Format	Top 40 Country
Network	Premiere
Program Director	Angie Fox	angief@kicks96.com

See **WJYW** (Union City).

WCTV TV (Cable-only channels 11, 20 & 21)

Address	2325 Chester Blvd., Richmond IN 47374
Telephone	765-973-8488
Fax	765-973-8489
E-mail	wctv@iue.edu
Web Site	www.wctv.info
On-air Hours	24/7
Broadcast Company	Whitewater Community Television
General Manager	John Schuerman	jschuerm@iue.edu
Accepts PSAs?	yes (contact John Dalton)
Notes	Channel formats: 11 (government); 20 (education); and 21 (public access). Non-commercial station. Does not broadcast local news.

WKOI TV
(Channels 43.1/TBN; 43.2/The Church Channel; 43.3/JCTV;43.4/Enlace USA-Spanish; & 43.5/Smile of a Child)

Mailing Address	P.O. Box 1057, Richmond IN 47375
Street Address	1702 S. 9th St., Richmond IN 47374
Telephone	765-935-2390
Fax	765-935-5367
Web Site	www.tbn.org
Network Affiliation	Trinity Broadcasting
On-air Hours	24/7
Broadcast Company	Trinity Broadcasting
General Manager	Marti Crick mcrick@tbn.org
Accepts PSAs?	yes—only from religious organizations (contact Marti Crick)
Notes	Non-commercial station. Does not broadcast local news.

RISING SUN

Ohio County

Ohio County News — Rising Sun Recorder

Mailing Address	P.O. Box 128, Rising Sun IN 47040
Street Address	235 Main St., Rising Sun IN 47040
Telephone	812-438-2011
Fax	812-438-3228
E-mail	risingsun@registerpublications.com
Publication Date	Weekly (Thursday)
Circulation	1,700-paid (combined)
Publishing Company	Register Publications
Publisher	Joe Awad editor@registerpublications.com
Editor	Tim Hillman risingsun@registerpublications.com
Advertising Manager	Loretta Day lday@registerpublications.com
Notes	Main office at Register Publications (Lawrenceburg).

ROCHESTER

Fulton County

Rochester Sentinel

Mailing Address	P.O. Box 260, Rochester IN 46975
Street Address	118 E. 8th St., Rochester IN 46975
Telephone	574-223-2111
Telephone (toll-free)	800-686-2112
Fax	574-223-5782
E-mail (news)	news@rochsent.com
Web Site	www.rochsent.com
Publication Date	Daily (Monday-Saturday)
Circulation	3,900-paid (Monday-Friday); 4,100-paid (Saturday)
Owner	Jack Overmyer
Publisher	Sarah O. Wilson show@rochsent.com
Editor	Bill Wilson wsw@rochsent.com
Advertising Manager	Karen Vojtasek karenv@rochsent.com
Sports Editor	Val Tsoutsouris valsports@rochsent.com
TMC/Shopper	The Compass (weekly)

Shopping Guide News

Mailing Address	P.O. Box 229, Rochester IN 46975
Street Address	617 Main St., Rochester IN 46975
Telephone	574-223-5417
Fax	574-223-8330
E-mail	shoppingguide@rtcol.com
Web Site	www.rtcol.com/~shoppingguide
Publication Date	Weekly (Wednesday)
Circulation	9,400 (free/delivered)
Owner/Publisher	Steve Foster
Owner/Advertising Mgr.	Barb Foster
Editor	Becky Melton

See **WQKV/K-Love** (listed under National Radio Stations).

WROI (92.1 FM)

Address	110 E. 8th St., Rochester IN 46975
Telephone	574-223-6059
Fax	574-223-2238
E-mail	wroi@rtcol.com
Web Site	www.wroifm.com
Wattage	4,300
Format	Oldies
Network Affiliations	ABC, Brownfield
On-air Hours	24/7
Broadcast Company	Bair Communications
General Manager	Tom Bair
News Director	Baron von Imhoof
Program Director	Cannon Fire
Sales Manager	Sue Bair
Accepts PSAs?	yes (contact Tom Bair)

ROCKPORT

Spencer County

Spencer County Journal-Democrat

Mailing Address	P.O. Box 6, Rockport IN 47635	
Street Address	541 Main St., Rockport IN 47635	
Telephone	812-649-9196	
Fax	812-649-9197	
E-mail	journal@psci.net	
Web Site	www.spencercountyjournal.com	
Publication Date	Weekly (Thursday)	
Circulation	6,100 (paid)	
Publishing Company	Landmark Community Newspapers	
Publisher	Kevin Lashbrook	publisher@perrycountynews.com
Editor	Vince Luecke	editor@spencercountyjournal.com
Advertising Manager	Theresa Lain	advertising@spencercountyjournal.com
Sports Editor	Brandon Cole	sports@spencercountyjournal.com
TMC/Shopper	The Lincoln's Country Shopper (weekly)	

ROCKVILLE

Parke County Sentinel
Parke County

Mailing Address	P.O. Box 187, Rockville IN 47872
Street Address	125 W. High St., Rockville IN 47872
Telephone	765-569-2033
Fax	765-569-1424
E-mail	sentinel@ticz.com
Web Site	www.parkecountysentinel.com
Publication Date	Weekly (Wednesday)
Circulation	4,400 (paid)
Publishing Company	Torch Newspapers
Publisher	Mary Jo Harney
General Manager	Kimberly White
Editor	Larry Bemis
Advertising Manager	Jane Kelp

See **WAXI** (Terre Haute).

ROYAL CENTER

Royal Centre Record
Cass County

Mailing Address	P.O. Box 638, Royal Center IN 46978
Street Address	102 S. Chicago St., Royal Center IN 46978
Telephone	574-643-3165
Fax	574-643-9440
E-mail (news)	editor@rcrecord.com
Publication Date	Weekly (Thursday)
Circulation	500 (paid)
Publishing Company	Small Town Publishing
Publisher/Advertising Mgr.	Laura Morrical
Editor	Rachel Carlson

RUSHVILLE

Rushville Republican
Rush County

Mailing Address	P.O. Box 189, Rushville IN 46173	
Street Address	126 S. Main St., Rushville IN 46173	
Telephone	765-932-2222	
Fax	765-932-4358	
Web Site	www.rushvillerepublican.com	
Publication Date	Semi-weekly (Tuesday, Thursday & Saturday)	
Circulation	3,100 (paid)	
Publishing Company	cnhi media	
Publisher	Laura Welborn	laura.welborn@indianamediagroup.com
Managing Editor	Kevin Green	kevin.green@indianamediagroup.com
Advertising Manager	Keith Wells	keith.wells@indianamediagroup.com
Sports Editor	Aaron Kirchoff	aaron.kirchoff@rushvillerepublican.com
TMC/Shopper	Extra (weekly)	

See **WENS** (Greenfield).

WIFE-FM (94.3 FM)

Address	102 N. Perkins St., Rushville IN 46173
Telephone	765-932-3983
Notes	Second office for WIFE-FM. Detailed information listed with the Connersville office.

See **WJCF** (Greenfield).

SALEM

Washington County

Salem Democrat — Salem Leader

Mailing Address	P.O. Box 506, Salem IN 47167	
Street Address	117-119 E. Walnut St., Salem IN 47167	
Telephone	812-883-3281	
Fax	812-883-4446	
E-mail (news)	stephanie@salemleader.com	
Web Site	www.salemleader.com	
Publication Date	Weekly: Thursday (Salem Democrat)	
	Weekly: Tuesday (Salem Leader)	
Circulation	6,500 (paid)	
Publishing Company	Leader Publishing Company of Salem, Inc.	
Publisher	Nancy Grossman	gm@salemleader.com
Editor	Stephanie Ferriell	stephanie@salemleader.com
Advertising Manager	Debbi Hayes	am@salemleader.com
TMC/Shopper	Your AD-Vantage (weekly)	

Washington County Edition—Bureau

Address	105 E. Walnut St., Salem IN 47167
Telephone	812-883-5555
Fax	812-883-3658
Notes	Main office in Pekin.

WSLM (1220 AM) — WSLM (97.9 FM) — WHAN TV (Cable-only channel 17)

Mailing Address	P.O. Box 385, Salem IN 47167
Street Address	1308 Hwy. 56 E., Salem IN 47167
Telephone	812-883-5750
Telephone (news)	812-883-3401
Fax	812-883-2797
Wattage	5,000 (WSLM AM)
	3,000 (WSLM FM)
Format	Farm/Country (WSLM AM)
	Talk/Gospel/Sports (WSLM FM)
On-air Hours	6:00 a.m. - 10:00 p.m. (radio)
	24/7 (television)
Broadcast Company	Community Broadcasting
General Mgr./Sales Mgr.	Don H. Martin
News Director	Rebecca White
Sports Director	Mark Abbott
Accepts PSAs?	yes (contact Rebecca White for radio and Don Martin for television)

See **WYGS** (Columbus).

SANTA CLAUS

See **WAXL** (Huntingburg).

SCOTTSBURG

Scott County Journal/Chronicle—Bureau

Mailing Address	P.O. Box 159, Scottsburg IN 47170
Street Address	183 E. McClain Ave., Scottsburg IN 47170
Telephone	812-752-3171
Telephone (toll-free)	800-254-9787
Fax	812-752-6486
E-mail	giveaway@c3bb.com
Web Site	www.gbpnews.com
Notes	Main office in Pekin.

WMPI (105.3 FM)

Mailing Address	P.O. Box 270, Scottsburg IN 47170	
Street Address	22 E. McClain Ave., Scottsburg IN 47170	
Telephone	812-752-3688	
Telephone (toll-free)	800-441-1053	
Telephone (news)	812-752-5612	
Fax	812-752-2345	
E-mail (news)	i1053news@verizon.net	
Web Site	www.i1053online.com	
Wattage	6,000	
Format	Country	
On-air Hours	24/7	
Broadcast Company	D. R. Rice Broadcasting	
General Mgr./Sales Mgr.	Tom Cull	wmpisales@verizon.net
News Director	Sharon Love	slove@i1053.com
Program/Promotions Dir.	John Ross	johnross3478@yahoo.com
Accepts PSAs?	yes (contact John Ross)	

SEYMOUR

Tribune

Mailing Address	P.O. Box 447, Seymour IN 47274	
Street Address	100 St. Louis Ave., Seymour IN 47274	
Telephone	812-522-4871	
Telephone (toll-free)	800-800-8212	
Telephone (news)	812-523-7051	
Fax	812-522-7691	
Fax (news)	812-522-3371	
E-mail (news)	dan_davis@link.freedom.com	
Web Site	www.tribtown.com	
Publication Date	Daily (Monday-Saturday)	
Circulation	8,500 (paid)	
Publishing Company	Freedom Communications, Inc.	
Publisher	Richard Davis	richard_davis@link.freedom.com
Editor	Dan Davis	dan_davis@link.freedom.com
Advertising Manager	Scott Embry	scott_embry@link.freedom.com
Sports Editor	Zach Spicer	zach_spicer@link.freedom.com
TMC/Shopper	The Tribune Marketplace Connection (weekly)	

See **WHUM** (Columbus).

WJAA (96.3 FM)

Address	1531 W. Tipton St., Seymour IN 47274
Telephone	812-523-3343
Telephone (toll-free)	877-523-3343
Fax	812-523-5116
E-mail	radio@wjaa.net
Web Site	www.wjaa.net
Wattage	3,000
Format	AAA/Alternative Rock
Network Affiliations	ABC, Westwood One
On-air Hours	24/7
Broadcast Company	Midland Media
General Mgr./News Dir.	Robert Becker robert@wjaa.net
Sales Manager	Tony Starkey tony@wjaa.net
Public Affairs Director	Kelly Landis kelly@wjaa.net
Accepts PSAs?	yes (contact Kelly Landis)

See **WJLR/K-Love** (listed under National Radio Stations).

WXKU (92.7 FM) — WZZB (1390 AM)

Mailing Address	P.O. Box 806, Seymour IN 47274
Street Address	1534 N. Ewing St., Seymour IN 47274
Telephone	812-522-1390
Fax	812-522-9541
On-air Hours	24/7
Broadcast Company	Traskom
General Manager	Blair Trask
News Director	Bud Shippee
Accepts PSAs?	yes (contact Jay Hubbard)

WXKU

E-mail	wxku@comcast.net
Web Site	www.kix92.com
Wattage	3,000
Format	Country
Network	ABC

WZZB

E-mail	wzzb@comcast.net
Wattage	1,000
Format	Soft Hits/Full Service
Networks	Jones, USA

SHELBYVILLE

Shelby County

Shelbyville News

Mailing Address	P.O. Box 750, Shelbyville IN 46176
Street Address	123 E. Washington St., Shelbyville IN 46176
Telephone	317-398-6631
Telephone (toll-free)	800-362-0114
Fax	317-398-0194
E-mail	shelbynews@shelbynews.com
Web Site	www.shelbynews.com
Publication Date	Daily (Monday-Saturday)
Circulation	9,400 (paid)
Publishing Company	Paxton Media Group
Publisher	Rachel Raney	rraney@shelbynews.com
Advertising Manager	Jody Street	jstreet@shelbynews.com
Sports Editor	Jeff Brown	jbrown@shelbynews.com
TMC/Shopper	The Extra (weekly)

See **WJCF** (Shelbyville).

WSVX (1520 AM) — WSVX (96.5 FM)

Address	2356 N. Morristown Rd., Shelbyville IN 46176
Telephone	317-398-2200
Telephone (toll-free)	866-270-WSVX
Fax	317-392-3292
E-mail	info@wsvx.com
E-mail (news)	news@wsvx.com
Web Site	www.wsvx.com
Wattage	1,000 (WSVX AM)
	165 (WSVX FM)
Format	Adult Hits
Network Affiliations	Brownfield, Network Indiana
On-air Hours	24/7
Broadcast Company	3 Towers Broadcasting
General Manager	Scott Huber	shuber@wsvx.com
News/Sports Director	Johnny McCrory	jmccrory@wsvx.com
Program/Public Affairs Dir.	Douglas Raab	draab@wsvx.com
Sales Manager	John Schoentrup	jschoentrup@wsvx.com
Accepts PSAs?	yes (contact Douglas Raab)
Notes	WSVX AM and WSVX FM are simulcast.

SHOALS

Martin County

The Shoals News

Mailing Address	P.O. Box 240, Shoals IN 47581
Street Address	311 High St., Shoals IN 47581
Telephone	812-247-2828
Fax	812-247-2243
Publication Date	Weekly (Wednesday)
Circulation	2,800 (paid)
Publisher/Editor	Steve Deckard	steve@theshoalsnews.com

SOUTH BEND

Saint Joseph County

South Bend Tribune

Address	225 W. Colfax Ave., South Bend IN 46626
Telephone	574-235-6161
Telephone (toll-free)	800-552-2795
Telephone (news)	574-235-6317
Fax (news)	574-236-1765
Fax (advertising)	574-239-2648
E-mail (news)	sbtnews@sbtinfo.com
Web Site	www.southbendtribune.com
Publication Date	Daily (Sunday-Saturday)
Circulation	57,000-paid (daily); 74,000-paid (Sunday)
Publishing Company	Schurz Communications
Publisher/Editor	David Ray	dray@sbtinfo.com
General Manager	Steven Funk	sfunk@sbtinfo.com
Managing Editor	Tim Harmon	tharmon@sbtinfo.com
Advertising Director	Carol Smith	csmith@sbtinfo.com
Sports Editor	Bill Bilinski	bbilinski@sbtinfo.com
Notes	Bureau in Mishawaka.

Tri-County News

Mailing Address	P.O. Box 6666, South Bend IN 46660
Telephone	574-243-4664
Fax	574-243-4916
E-mail	admin@tricountynewsinc.com
Web Site	www.tricountynewsinc.com
Publication Date	Weekly (Friday)
Circulation	1,000 (paid)
Owner/Managing Editor	Cherie Jolly
Office Manager	Lisa Andrysiak

See **Tribune Business Weekly** (listed under Business/Specialty Publications).

WDND (1490 AM) — **WNDV** (92.9 FM) — **WSMM** (102.3 FM) — **WZOW** (97.7 FM)

Address	3371 W. Cleveland Rd., STE 300, South Bend IN 46628
Telephone	574-273-9300
Fax	574-273-9090
On-air Hours	24/7
Broadcast Company	Artistic Media Partners
General Manager	Arthur Angotti III arthur@artisticradio.com
News Director	Karen Rite karen@u93.com
Promotions Director	A. J. Seliga aj@u93.com
Accepts PSAs?	yes (contact Karen Rite)

WDND

Web Site	www.espnradio1490.com
Wattage	1,000
Format	Sports
Network	ESPN
Program Director	Bob Henning rhenn15090@aol.com
Sales Manager	Dick O'Day dick@wzow.com

WNDV

Web Site	www.u93.com
Wattage	20,000
Format	Adult Contemporary Hit Radio
Network	Westwood One
Program Director	Karen Rite karen@u93.com
Sales Manager	Pam Homan pam@u93.com

WSMM

Wattage	3,000
Format	Adult Contemporary
Network	Westwood One
Program Director	Karen Rite karen@u93.com
Sales Manager	Pam Homan pam@u93.com

WZOW

Web Site	www.wzow.com
Wattage	6,000
Format	Rock
Network	Westwood One
Program Director	Karen Rite karen@u93.com
Sales Manager	Dick O'Day dick@wzow.com
Notes	Serves Elkhart & South Bend.

WETL (91.7 FM)

Address	1902 S. Fellows, South Bend IN 46613
Telephone	574-283-8432
Fax	574-283-8405
E-mail	jovermyer@sbcsc.k12.in.us
Wattage	3,000
Format	Educational
On-air Hours	24/7
Owner	South Bend Community School Corp.
News/Program Director	John Overmyer jovermyer@sbcsc.k12.in.us
Accepts PSAs?	yes
Notes	Non-commercial station.

See **WFRN** (Elkhart).

WHME (103.1 FM) — WHPD (92.1 FM) — WHPZ (96.9 FM)

Address	61300 S. Ironwood Rd., South Bend IN 46614
Telephone	574-291-8200
Fax	574-291-9043
E-mail	pulse@lesea.com
On-air Hours	24/7
Broadcast Company	LeSea Broadcasting Corp.
General Manager	Peter Sumrall
Program Director	Gary Hegland ghegland@lesea.com
Sports Director	Chuck Freeby cfreeby@lesea.com
Sales Manager	Anna Riblet ariblet@lesea.com
Public Affairs Director	Angela Sumrall asumrall@lesea.com
Accepts PSAs?	yes (contact Kimberly Ann, pulse@lesea.com)

WHME
Web Site	www.lesea.com/whmefm/
Wattage	3,000
Format	Christian Talk/Teaching
Notes	Does not broadcast local news.

WHPD & WHPZ
Web Site	www.pulsefm.com
Wattage	5,000 (WHPD)
	3,000 (WHPZ)
Format	Contemporary Christian
Notes	WHPZ and WHPD are simulcast.

WSBL-LP (98.1 FM & 106.5 FM)

Mailing Address	P.O. Box 402, South Bend IN 46624
Street Address	2015 W. Western Ave., STE 136, South Bend IN 46619
Telephone/Fax	574-232-3212
E-mail	wsbllp@sbcglobal.net
Web Site	www.radiosaborlatino.com
Wattage	100
Format	Public Radio/Hispanic
On-air Hours	24/7
Broadcast Company	Radio Sabor Latino
General Manager	Edwin Gonzalez
Director	Eliud Villanueva
Accepts PSAs?	yes (contact Edwin Gonzalez)
Notes	Non-commercial station. 98.1 FM serves South Bend. 106.5 FM serves Goshen.

See **WSBT** (Mishawaka).

WUBS (89.7 FM)

Mailing Address	P.O. Box 3931, South Bend IN 46619
Street Address	702 Lincolnway West, South Bend IN 46616
Telephone	574-287-1266
Fax	574-287-2478
E-mail	revwilliams@wubs.org
Web Site	www.wubs.org
Wattage	1,500
Format	Inspirational
On-air Hours	24/7
Broadcast Company	Interface Christian Union
Owner/General Manager	Rev. Sylvester Williams, Jr.
Program Director	Shane Williams
Accepts PSAs?	yes
Notes	Non-commercial station.

WUBU (106.3 FM)

Address	401 E. Colfax Ave., STE 300, South Bend IN 46617	
Telephone	574-233-3505	
Fax	574-233-0580	
E-mail (news)	news@michiananewschannel.com	
Web Site	www.wubufm.com	
Wattage	3,000	
Format	Urban Adult Contemporary	
Network Affiliations	ABC	
On-air Hours	24/7	
Broadcast Company	Partnership Radio	
General Manager	Abe Thompson	
News/Program Director	Gene Walker	gwalker@wubufm.com
Sales/Promotions Manager	Erica Morse	emorse@wubufm.com
Public Affairs Director	Vivian Sallie	
Accepts PSAs?	yes (contact Gene Walker)	

See **WZOC** (Mishawaka).

WHME TV (Channels 46.1/main programming & 46.2/World Harvest Television)
WHNW TV (Channel 18)

Address	61300 S. Ironwood Rd., South Bend IN 46614
Telephone	574-291-8200
Fax	574-291-9043
E-mail	whme@lesea.com
Web Site	www.whme.com
Network Affiliation	independent
On-air Hours	24/7
Broadcast Company	LeSea Broadcasting
General Manager	Peter Sumrall
Sports Director	Chuck Freeby
General Sales Manager	Anna Riblet
Program Director	Colleen Halt
Accepts PSAs?	yes (contact Colleen Halt)
Notes	WHNW TV serves Gary and is a simulcast of World Harvest Television.

WNDU TV (Channel 16.1)

Mailing Address	P.O. Box 1616, South Bend IN 46634
Street Address	54516 State Road 933, South Bend IN 46637
Telephone	574-284-3000
Telephone (news)	574-284-3016
Telephone (toll-free news)	800-631-6397
Fax	574-284-3009
Fax (news)	574-284-3022
Fax (sales)	574-284-3115
E-mail (news)	newscenter16@wndu.com
Web Site	www.wndu.com
Network Affiliation	NBC
On-air Hours	24/7
Broadcast Company	Gray Television
General Manager	John O'Brien	john.obrien@wndu.com
News Director	C. J. Beutien	cj.beutien@wndu.com
News Assignment Editor	Mike Pease	mike.pease@wndu.com
Sports Director	Jeff Jeffers	jeff.jeffers@wndu.com
General Sales Manager	Howard Voss	howard.voss@wndu.com
Local Sales Manager	Michael German	michael.german@wndu.com
Program/Public Affairs Dir.	Michael Fowler	michael.fowler@wndu.com
Accepts PSAs?	yes (contact Michael Fowler)	

WSJV TV (Channel 28.1)

Mailing Address	P.O. Box 28, South Bend IN 46624
Street Address	58096 County Road 7 S., Elkhart IN 46517
Telephone	574-679-9758
Telephone (toll-free)	800-975-8881
Fax	574-294-1267
Fax (news)	574-522-7609
Fax (sales)	574-294-1324
E-mail	fox28@fox28.com
E-mail (news)	fox28news@fox28.com
Web Site	www.fox28.com
Network Affiliation	Fox
On-air Hours	24/7
Broadcast Company	Quincy Newspapers, Inc.
News Director	Ed Kral	ekral@fox28.com
News Assignment Editor	Jeff Amos	jamos@fox28.com
Sports Director	Dean Huppert	dhuppert@fox28.com
General Sales Manager	Mike Leyes	mleyes@fox28.com
National Sales Manager	Dave Lenaway	dlenaway@fox28.com
Program Director	Heather Stewart	hstewart@fox28.com
Promotions Director	Paul Wasowski	pwasowski@fox28.com
Accepts PSAs?	yes (contact fox28@fox28.com)	

SOUTH WHITLEY

Whitley County

Tribune-News

Address	113 S. State St., South Whitley IN 46787
Telephone/Fax	260-723-4771
E-mail	email@tribunenews.biz
Publication Date	Weekly (Wednesday)
Circulation	1,200 (paid)
Publishing Company	Smith Publications
Publisher/Editor	Teresa Smith
General Manager	Tonya Porter
Advertising Manager	Linda Hoskins

SPEEDWAY

Marion County

Speedway Town Press — West Side Messenger

Address	1538 Main St., Speedway IN 46224
Telephone	317-241-4345
Fax	317-241-4386
E-mail	thepress@in-motion.net
Publication Date	Weekly (Wednesday)
Circulation	5,500 (free/delivered)
Publishing Company	Speedway Northwest, Inc.
Publisher/Editor	Beth Sullivan
General Manager	Barbara Perkins
Advertising Manager	Shirley Nelson
Notes	West Side Messenger serves west side of Indianapolis.

SPENCER

Owen County

Owen Leader — Spencer Evening World

Mailing Address	P.O. Box 226, Spencer IN 47460
Street Address	114 E. Franklin St., Spencer IN 47460
Telephone	812-829-2255
Fax	812-829-4666
E-mail (news)	editor@spencereveningworld.com
Web Site	www.spencereveningworld.com
Publication Date	Weekly: Thursday (Owen Leader)
	Daily: Monday-Friday (Spencer Evening World)
Circulation	300-paid (Owen Leader)
	3,700-paid (Spencer Evening World)
Publishing Company	Spencer Evening World Inc.
General Manager	John A. Gillaspy
Editor	Travis Curry editor@spencereveningworld.com

SULLIVAN

Sullivan County

Sullivan Daily Times

Mailing Address	P.O. Box 130, Sullivan IN 47882
Street Address	115 W. Jackson St., Sullivan IN 47882
Telephone	812-268-6356
Telephone (toll-free)	800-264-6356
Fax	812-268-3110
E-mail (news)	tomgettinger@sullivan-times.com
Web Site	www.sullivan-times.com
Publication Date	Daily (Monday-Friday)
Circulation	5,000 (paid)
Publisher	Nancy Gettinger	publisher@sullivan-times.com
Editor	Pete Wilson	editor@sullivan-times.com
Managing Editor	Tom Gettinger	tomgettinger@sullivan-times.com

WNDI (1550 AM) — WNDI (95.3 FM)

Address	556 E. State Road 54, Sullivan IN 47882
Telephone	812-268-6322
Fax	812-268-6652
E-mail	wndi@joink.com
Wattage	250 (WNDI AM)
	6,000 (WNDI FM)
Format	Today's Country
On-air Hours	24/7
Broadcast Company	JTM Broadcasting Corp.
General Manager	John Montgomery
Accepts PSAs?	yes
Notes	WNDI AM & WNDI FM are simulcast.

SYRACUSE

Kosciusko County

'the Paper'—Bureau

Address	102 E. Main St., Syracuse IN 46567
Telephone	574-457-3666
Fax	574-457-3852
E-mail	syracuse@the-papers.com
Office Manager	Shawna VanLue
Notes	Main office in Milford.

TELL CITY

Perry County News
Perry County

Mailing Address	P.O. Box 309, Tell City IN 47586
Street Address	537 Main St., Tell City IN 47586
Telephone	812-547-3424
Fax	812-547-2847
E-mail	editor@perrycountynews.com
Web Site	www.perrycountynews.com
Publication Date	Semi-weekly (Monday & Thursday)
Circulation	6,800 (paid)
Publishing Company	Landmark Community Newspapers
Publisher	Kevin Lashbrook publisher@perrycountynews.com
Editor	Vince Luecke editor@perrycountynews.com
Advertising Manager	Cindy Dauby advertising @perrycountynews.com
Sports Editor	Larry Goffinet sports@perrycountynews.com
TMC/Shopper	Lincolnland Shopping Guide (weekly)

WLME (102.7 FM) — WTCJ (1230 AM) — WTCJ (105.7 FM)

Address	1115 Tamarack Rd., STE 500, Owensboro KY 42301
Telephone	270-683-5200
Telephone (toll-free)	800-547-8121
Fax	270-688-0108
E-mail (news)	mchaney@cromwellradio.com
Web Site	www.tellcityradio.com
Wattage	25,000 (WLME)
	1,000 (WTCJ AM)
	6,000 (WTCJ FM)
Format	Hot Adult Contemporary (WLME)
	Soft Adult Contemporary (WTCJ AM)
	Classic Hits (WTCJ FM)
Network Affiliations	ABC, Dial Global
On-air Hours	24/7
Broadcast Company	Hancock Communications
General Mgr./Sales Mgr.	Lee Wilson lwilson@cromwellradio.com
News Director	Mike Chaney mchaney@cromwellradio.com
Program//Sports Director	Jeff Morgan jmorgan@cromwellradio.com
Public Affairs/Promotions	Derik Hancock dhancock@cromwellradio.com
Accepts PSAs?	yes (contact Derik Hancock)
Notes	Stations serve Tell City.

TERRE HAUTE

Vigo County

See **Family Life** (listed under Parenting/Specialty Publications).

See **Indiana Statesman** (listed under College Campus/Specialty Publications).

Tribune-Star
Mailing Address	P.O. Box 149, Terre Haute IN 47808
Street Address	222 S. 7th St., Terre Haute IN 47807
Telephone	812-231-4200
Telephone (toll-free)	800-783-8742
Fax	812-231-4347
Fax (news)	812-231-4321
Fax (advertising)	812-231-4234
E-mail	community@tribstar.com
Web Site	www.tribstar.com
Publication Date	Daily (Sunday-Saturday)
Circulation	27,000 (paid)
Publishing Company	Community Newspaper Holdings Inc.
Publisher	B. J. Riley — bjriley@tribstar.com
Editor	Max Jones — max.jones@tribstar.com
Advertising Manager	Tanya Wilhoyte — tanya.wilhoyte@tribstar.com
Sports Editor	Todd Golden — todd.golden@tribstar.com

See **Wabash Valley Journal of Business** (listed under Business/Specialty Publications).

WAXI (104.9 FM) — WBOW (1300 AM) — WBOW (102.7 FM) — WSDM (92.7 FM) WSDX (1130 AM)

Address	1301 Ohio St., Terre Haute IN 47807
Telephone	812-234-9770
Telephone (toll-free)	800-939-9770
Fax	812-238-1576
On-air Hours	24/7
Broadcast Company	Crossroads Communications
General Manager	Mike Peterson mpeter@wsdm.com
Sales Manager	Robin Woods
Accepts PSAs?	yes (contact Bill Cook)
Notes	Stations do not broadcast local news.

WAXI

Web Site	www.waxifm.com
Wattage	3,000
Format	Gold Hits (60s & 70s)
Network	X-Radio (Mike Harvey Super Gold)
Program Director	Marty Combs partymarty@crock927.com
Notes	Serves Rockville.

WBOW AM & WSDX

Web Site	www.espnsportsradio.com
Wattage	500
Format	Sports
Network	ESPN
Program Director	Adam Michaels adam@radioworksforme.com
Notes	WBOW AM & WSDX are simulcast. WSDX serves Brazil.

WBOW FM

Web Site	www.wbowfm.com
Wattage	50,000
Format	Hot Adult Contemporary
Network	Dial Global
Program Director	Adam Michaels adam@radioworksforme.com

WSDM

Web Site	www.crock927.com
Wattage	6,000
Format	Country & Rock
Network	Dial Global
Program Director	Marty Combs partymarty@crock927.com
Notes	Serves Brazil.

WCRT (88.5 FM)

Address	2108 W. Springfield Ave., Champaign IL 61821
Telephone	217-359-8232
Telephone (toll-free)	866-917-9245
Fax	217-359-7374
E-mail	wbgl@wbgl.org
Web Site	www.wbgl.org
Wattage	500
Format	Adult Contemporary Christian
On-air Hours	24/7
Owner	Illlinois Bible Institute
General Manager	Jeff Scott
News Director	Tim Sinclair
Program Director	Ryan Springer
Sales Manager	Zoe Fuller
Public Affairs Director	Ginee Scott
Promotions Director	Jennifer Briski
Accepts PSAs?	yes (contact Ginee Scott (familyplanner@wbgl.org)
Notes	WCRT serves Terre Haute and is a simulcast of WBGL (91.7 FM) in Champaign, IL. Non-commercial station.

WHOJ (91.9 FM)

Address	4424 Hampton Ave., St. Louis MO 63109
Telephone	314-752-7000
Telephone (toll-free)	877-305-1234
Fax	314-752-7702
E-mail	covenantnetwork@juno.com
Web Site	www.covenantnet.net.
Wattage	1,500
Format	Catholic
Network Affiliations	EWTN
On-air Hours	24/7
Broadcast Company	Covenant Network
General Manager	Tony Holman
Accepts PSAs?	yes (contact Tony Holman)
Notes	Non-commercial station. Serves Terre Haute.

WIBQ (98.5 FM) — WMGI (100.7 FM) — WWSY (95.9 FM)

Address	824 S. 3rd St., Terre Haute IN 47807
Telephone	812-232-4161
Fax	812-234-9999
On-air Hours	24/7
Broadcast Company	Midwest Communications
General Manager	Becky Dole beckyd@mwcradio.com
News Director	Steve Hall steve@wibqfm.com
Sales Manager	Dave Vanlandingham dave@1007mixfm.com
Public Affairs/Promotions	Stacey Kaye stacey@1007mixfm.com
Notes	Also operates WPRS AM, which serves Paris, IL.

WIBQ

Web Site	www.wibqfm.com
Wattage	50,000
Format	News/Talk
Program Director	Bill Cain bill.cain@mwcradio.com
Accepts PSAs?	yes (contact Bill Cain)

WMGI

Web Site	www.1007mixfm.com
Wattage	50,000
Format	Contemporary Hit Radio
Program Director	Bill Cain bill.cain@mwcradio.com
Accepts PSAs?	yes (contact Bill Cain)

WWSY

E-mail	info@959thevalley.com
Web Site	www.959thevalley.com
Wattage	4,100
Format	Adult Hits
Program Director	Natalie Randall natalie@959thevalley.com
Accepts PSAs?	yes (contact Natalie Randall)

WISU (89.7 FM)

Address	217 Dreiser Hall, Indiana State U., Terre Haute IN 47809
Telephone	812-237-3248
Telephone (news)	812-237-3690
Fax	812-237-3241
E-mail	wisufm@indstate.edu
Web Site	www.wisu.org
Wattage	13,500
Format	Alternative Rock/Urban
On-air Hours	24/7
Owner	Indiana State University
General Manager	Dave Sabaini
Accepts PSAs?	yes
Notes	Non-commercial station.

WMHD (90.7 FM)

Address	5500 Wabash Ave., Terre Haute IN 47803
Telephone	812-872-6923
Fax	812-872-6926
E-mail	manager@wmhdradio.org
Web Site	www.wmhdradio.org
Wattage	1,400
Format	College/Free Format
On-air Hours	24/7
Owner	Rose-Hulman Institute of Technology
Faculty Advisor	Kevin Lanke lanke@rose-hulman.edu
Accepts PSAs?	yes
Notes	Non-commercial station.

WPFR (1480 AM) — WPFR (93.9 FM)

Address	18889 N. 2350 St., Dennison IL 62423
Telephone	217-826-9673
Telephone (toll-free)	877-939-1480
E-mail	wpfr@joink.com
Web Site	www.wordpower.us
Wattage	5,000-day; 1,000-night (WPFR AM)
	2,350 (WPFRM FM)
Format	Christian
Network Affiliations	Moody
On-air Hours	24/7
Broadcast Company	Word Power, Inc.
General Manager	Paul Ford
Program Director	Eleanor Ford
Accepts PSAs?	yes
Notes	WPFR AM serves Terre Haute. WPFR FM serves Clinton.

See **WQTY** (Vincennes).

WTHI (99.9 FM) — **WWVR** (105.5 FM)

Mailing Address	P.O. Box 1486, Terre Haute IN 47808
Street Address	918 Ohio St., Terre Haute IN 47807
Telephone	812-232-9481
Fax	812-234-0089
E-mail (news)	frush@wthi.emmis.com
On-air Hours	24/7
Broadcast Company	Emmis Communications
General Manager	James Conner — jconner@wthi.emmis.com
News Director	Frank Rush — frush@wthi.emmis.com
Public Affairs/Promotions	Amy Clark — aclark@wthi.emmis.com
Accepts PSAs?	yes (contact Amy Clark)

WTHI

Web Site	www.hi99.com
Wattage	50,000
Format	Country
Network	ABC Entertainment
Program Director	Barry Kent — bkent@wthi.emmis.com

WWVR

Web Site	www.1055theriver.com
Wattage	3,000
Format	Rock
Network	Bob & Tom
Program Director	Ed Zeppelin — ezeppelin@wwvr.emmis.com

WFXW TV (Channel 39.1) — **WTWO TV** (Channels 2.1 & 36.1)

Mailing Address	P.O. Box 9268, Terre Haute IN 47808
Street Address	10849 N. U.S. Hwy. 41, Farmersburg IN 47850
Fax	812-696-2755
Fax (news)	812-696-2000
Fax (sales)	812-696-2718
E-mail	tsturgess@wtwo.com
Web Site	www.mywabashvalley.com
On-air Hours	24/7
General Manager	Tim Sturgess — tsturgess@wtwo.com
News Director	Tom McClanahan — news@wtwo.com
News Assignment Editor	Wendell Hudson — news@wtwo.com
Sports Director	Jason Pensky — jason@aol.com
General Sales Manager	Jeremiah Turner — jturner@wtwo.com
New Media Manager	Tim Hennessy — thennessy@wtwo.com
Accepts PSAs?	yes (contact Tim Sanders, tsanders@wtwo.com)

WFXW TV

Telephone	812-238-3838
Network	Fox
Broadcast Company	Mission Broadcasting
Station Manager	Lois Mathes — business@wfxw.tv

WTWO TV

Telephone	812-696-2121
E-mail	station@wtwo.com
Network	NBC
Broadcast Company	Nexstar Broadcasting
Program Director	Tim Sturgess — tsturgess@wtwo.com

WTHI TV (Channel 10)

Mailing Address	P.O. Box 9606, Terre Haute IN 47808
Street Address	918 Ohio St., Terre Haute IN 47807
Telephone	812-232-9481
Telephone (toll-free)	800-589-8810
Telephone (news)	812-232-4953
Fax	812-232-8953
Fax (news)	812-232-3694
E-mail (news)	news10@wthitv.com
Web Site	www.wthitv.com
Network Affiliation	CBS
On-air Hours	24/7
Broadcast Company	LIN Television
General Manager	Todd Weber	tweber@wthitv.com
News Director	Susan Dinkel	sdinkel@wthitv.com
News Assignment Editor	Mandi Scott	ascott@wthitv.com
Sports Director	Rick Semmler	rsemmler@wthitv.com
General Sales Manager	Nick Telezyn	ntelezyn@wthitv.com
Local Sales Manager	Jim Swander	jswander@wthitv.com
Program Director	Rod Garvin	rgarvin@wthitv.com
Public Affairs Director	Chris Wood	cowood@wthitv.com
Promotions Director	David Shearer	dshearer@wthitv.com
Accepts PSAs?	yes (contact Chris Wood)

TIPTON

Tipton County

Tipton County Tribune

Mailing Address	P.O. Box 248, Tipton IN 46072
Street Address	116 S. Main St., STE A, Tipton IN 46072
Telephone	765-675-2115
Fax	765-675-4147
E-mail	tiptontri@netscape.net
E-mail (news)	tiptoneditor@elwoodpublishing.com
Web Site	www.elwoodpublishing.com
Publication Date	Daily (Monday-Saturday)
Circulation	2,600 (paid)
Publishing Company	Elwood Publishing Co.
Publisher	Robert Nash	elpub@elwoodpublishing.com
Editor	Jackie Henry	tiptoneditor@elwoodpublishing.com
Advertising Manager	Mike Brown	tiptonads@elwoodpublishing.com
Sports Editor	Michelle Garmon	tiptonsports@elwoodpublishing.com
TMC/Shopper	Leader-Tribune Review West (weekly)

UNION CITY

WJYW (88.9 FM)   Randolph County

Mailing Address	P.O. Box 445, Union City IN 47390
Street Address	505 S. Division St., Union City OH 45390
Telephone	937-968-5633
Telephone (toll-free)	877-335-4569
Fax	937-968-3320
E-mail	office@889joyfm.com
Web Site	www.889joyfm.com
Wattage	4,100
Format	Contemporary Christian Music
Network Affiliations	Salem
On-air Hours	24/7
Broadcast Company	Positive Alternative Radio, Inc.
General Manager	Dan Franks dan@889joyfm.com
Accepts PSAs?	yes (contact Dan Franks)
Notes	Non-commercial station. Repeats on 94.5 FM (Richmond).

UPLAND

WTUR (89.7 FM)   Grant County

Address	236 W. Reade Ave., Upland IN 46989
Telephone	765-998-5263
Fax	765-998-4810
E-mail	wtur@taylor.edu
Wattage	150
Format	Christian Rock
Owner	Taylor University
General Manager	Kathy Bruner ktbruner@taylor.edu
Accepts PSAs?	yes
Notes	Non-commercial station.

VALPARAISO

The Chronicle   Porter County

Address	208 Elm St., Valparaiso IN 46383
Telephone	219-462-1488
Fax	219-462-3897
E-mail	chronicle@greatlakesmarketing.net
Publication Date	Weekly (Wednesday)
Circulation	28,500 (free/newsstand & mailed)
Publishing Company	Great Lakes Media, Inc.
Publisher	Brenda Kleihege brenda.graphics@verizon.net
General Manager	Tom Sanders tsanders-chronicle@comcast.net
Editor	Deb Johnson chronicle@greatlakesmarketing.net
Office Manager	Molly Tanner
Notes	Publishes 4 editions serving Chesterton, Hobart, Portage and Valparaiso.

Panorama Magazine — The WRite Stuff

Address	656 S. Franklin St., Valparaiso IN 46383
Telephone/Fax	219-464-9237
Telephone (toll-free)	800-359-2845
E-mail	baxterdesign@comcast.net
Publication Date	Monthly
Publisher	Sue Baxter

Panorama Magazine

E-mail (news)	panoramanow@comcast.net
Web Site	www.panoramanow.com
Circulation	35,000 (free/mailed & deliverd)
Publishing Company	Baxter Design & Advertising
Editor	Sue Baxter
Advertising Manager	John Sabo

The WRite Stuff

E-mail (news)	puccini99@aol.com
Web Site	www.whitingindiana.com/about_the_chamber.php
Circulation	6,100 (free to all homes & businesses in Whiting & Robertsdale)
Publishing Company	The Whiting Robertsdale Chamber of Commerce
Editor	Gayle Kosalko
Advertising Manager	Sue Baxter
Chamber Director	MaryLu Gregor
Notes	Serves Whiting and Robertsdale.

The Times of Northwest Indiana—Bureau

Address	1111 Glendale Blvd., Valparaiso IN 46383	
Telephone	219-462-5151	
Fax	219-465-7298	
Porter County Editor	John Scheibel	jscheibel@nwitimes.com
Notes	Main office in Munster.	

See **Torch** (listed under College Campus/Specialty Publications).

WAKE (1500 AM) — WLJE (105.5 FM) — WXRD (103.9 FM) — WZVN (107.1 FM)

Address	2755 Sager Rd., Valparaiso IN 46383
Telephone	219-462-6111
Fax	219-462-4880
E-mail	scottr@radiooneindiana.com
E-mail (news)	laura@radiooneindiana.com
On-air Hours	24/7
Broadcast Company	Radio One Communications
General Manager	Leigh Ellis lellis@radiooneindiana.com
Operations/Promotions Dir.	Scott Rosenberg scottr@radiooneindiana.com
News/Public Affairs Dir.	Laura Waluszko laura@radiooneindiana.com
Program Director	Don Clark donclark@radiooneindiana.com
Sales Manager	O. J. Jackson ojj@radiooneindiana.com
Accepts PSAs?	yes (contact Laura Waluszko)

WAKE

Web Site	www.wakeradio.com
Wattage	1,000
Format	Adult Pop Standards

WLJE

Web Site	www.indiana105.com
Wattage	3,000
Format	Country

WXRD

Web Site	www.xrock1039.com
Wattage	3,000
Format	Classic Rock
Network	Bob & Tom

WZVN

Web Site	www.z1071.com
Wattage	3,000
Format	Adult Hits

WCJL (90.9 FM) — WHLP (89.9 FM) — WJCO (91.3 FM) — WJCY (91.5 FM) WOJC (89.7 FM) — WQKO (91.9 FM) — WTMK (88.5 FM)

Address	150 W. Lincolnway, STE 2001, Valparaiso IN 46383
Telephone	219-548-5800
Telephone (toll-free)	866-303-9457
Fax	219-548-5808
E-mail	info@calvaryradionetwork.com
Web Site	www.calvaryradionetwork.com
Wattage	1,000 (WCJL)
	8,000 (WHLP)
	350 (WJCO)
	475 (WJCY)
	3,000 (WOJC & WQKO)
	1,500 (WTMK)
Format	Praise/Worship/Teaching
Network Affiliations	Calvary Radio Network
On-air Hours	24/7
Broadcast Company	Calvary Chapel Costa Mesa
General Manager	Jim Motshagen jmots@calvaryradionet.com
Program Director	Kathy Motshagen kmots@calvaryradionet.com
Public Affairs Director	Josh McAfee joshmcafee@calvaryradionet.com
Accepts PSAs?	yes (contact info@calvaryradionetwork.com)
Notes	Non-commercial stations. Stations do not broadcast local news. Service areas: Bloomington/Morgantown (WCJL), Valparaiso (WHLP), Montpelier (WJCO), Cicero (WJCY), Crothersville (WOJC), Howe (WQKO), and Lowell (WTMK). Also has translators within the Chicago metro area.

See **WFRN** (Elkhart).

WVLP (98.3 FM)

Address	256 W. Indiana Ave., Valparaiso IN 46383
Telephone	219-476-9000
E-mail	info@wvlp.org
Web Site	www.wvlp.org
Wattage	100
Format	Community Radio
On-air Hours	24/7
Broadcast Company	Neighbors Broadcasting
General Manager	Gregg Kovach info@wvlp.org
Accepts PSAs?	yes (contact Gregg Kovach)
Notes	Non-commercial station.

WVUR (95.1 FM)

Address	Valparaiso U., 1809 Chapel Dr., 32 Schnabel Hall, Valparaiso IN 46383
Telephone	219-464-5383
Telephone (studio)	219-464-6673
E-mail	thesource95.1@valpo.edu
Web Site	www.valpo.edu/wvur
Wattage	36
Format	Modern Rock
On-air Hours	24/7
Owner	Valparaiso University
Faculty Advisor	Paul Oren paul.oren@valpo.edu
Accepts PSAs?	yes
Notes	Non-commercial station.

VERSAILLES

Osgood Journal — Versailles Republican
Ripley County

Mailing Address	P.O. Box 158, Versailles IN 47042
Street Address	115 S. Washington St., Versailles IN 47042
Telephone	812-689-6364
Fax	812-689-6508
E-mail	publication@ripleynews.com
Web Site	www.ripleynews.com
Publication Date	Weekly: Tuesday (Osgood Journal)
	Weekly: Thursday (Versailles Republican)
Circulation	5,200 (paid)
Publishing Company	Ripley Publishing Co.
Publisher	Linda Chandler lchandler@ripleynews.com
Editor	Wanda Burnett wburnett@ripleynews.com
TMC/Shopper	Spotlight Advertiser (weekly)

See **WKRY** (Columbus).

VEVAY

Switzerland Democrat — Vevay Reveille-Enterprise
Switzerland County

Mailing Address	P.O. Box 157, Vevay IN 47043
Street Address	111 W. Market St., Vevay IN 47043
Telephone	812-427-2311
Telephone (toll-free)	877-848-3829
Fax	812-427-2793
E-mail	news@vevaynewspapers.com
Web Site	www.vevaynewspapers.com
Publication Date	Weekly (Thursday)
Circulation	4,000-paid (combined)
Publishing Company	Vevay Newspapers, Inc.
Publisher	Don Wallis Jr.
Editor	Patrick Lanman
Advertising Manager	Erin Williams

WKID (95.9 FM)

Address	118 W. Main St., Vevay IN 47043
Telephone	812-427-9590
Telephone (toll-free)	888-959-9543
Fax	812-427-2492
E-mail (news)	news@k959froggy.com
Web Site	www.k959froggy.com
Wattage	3,000
Format	Country
Network Affiliations	Network Indiana
On-air Hours	24/7
Broadcast Company	Dial Broadcasting Inc.
General Mgr./Program Dir.	Ken Trimble ken@k959froggy.com
News Director	Mike Wigren news@k959froggy.com
Sales Manager	Helen Peelman helen@k959froggy.com
Accepts PSAs?	no

VINCENNES

Knox County

Vincennes Sun-Commercial

Mailing Address	P.O. Box 396, Vincennes IN 47591
Street Address	702 Main St., Vincennes IN 47591
Telephone	812-886-9955
Telephone (toll-free)	800-876-9955
Fax (news)	812-885-2235
Fax (advertising)	812-885-2237
E-mail (news)	vscnews@suncommercial.com
Web Site	www.suncommercial.com
Publication Date	Daily (Sunday-Friday)
Circulation	9,500-paid (daily); 11,600-paid (Sunday)
Publishing Company	Paxton Media Group
Publisher	Vickie Palmer
Managing Editor	Gayle Robbins
Sports Editor	David Staver

WAOV (1450 AM) — **WBTO** (102.3 FM) — **WQTY** (93.3 FM) — **WRCY** (1590 AM) **WREB** (94.3 FM) — **WUZR** (105.7 FM) — **WWBL** (106.5 FM) — **WYFX** (106.7 FM) **WZDM** (92.1 FM)

Mailing Address	P.O. Box 242, Vincennes IN 47591
Street Address	522 Busseron St., Vincennes IN 47591
Telephone	812-882-6060
Fax	812-885-2604
Fax (sales)	812-882-7770
E-mail	marklange@originalcompany.com
E-mail (news)	news@originalcompany.com
On-air Hours	24/7
Broadcast Company	The Original Company, Inc.
General Manager	Mark Lange marklange@originalcompany.com
Sports Director	Dave Young daveyoung@originalcompany.com
Sales Manager	Michelle York michelleyork@originalcompany.com
Promotions Director	Ashley Cantwell ashleycantwell@originalcompany.com
Accepts PSAs?	no

WAOV

Web Site	www.waovam.com
Wattage	1,000
Format	News/Talk/Sports
News Director	Tom Lee tomlee@originalcompany.com
Program Director	Jonathan Lange jonathanlange@originalcompany.com

WBTO

Web Site	www.wbtofm.com
Wattage	3,000
Format	Classic Rock
News Director	John Szink johnszink@originalcompany.com
Program Director	Jonathan Lange jonathanlange@originalcompany.com
Notes	Serves Petersburg and Jasper.

WQTY

Web Site	www.wqtyradio.com
Wattage	25,000
Format	Oldies
News Director	John Szink johnszink@originalcompany.com
Program Director	Jonathan Lange jonathanlange@originalcompany.com
Notes	Serves Linton and Terre Haute.

WRCY

Web Site	www.wrcyam.com
Wattage	500
Format	Country
Program Director	Jonathan Lange jonathanlange@originalcompany.com
Notes	Serves Mount Vernon and has studio there.

WREB

Web Site	www.wrebfm.com
Wattage	3,000
Format	Adult Contemporary
Program Director	Jonathan Lange jonathanlange@originalcompany.com
Notes	Serves Greencastle and has studio there.

WUZR

Web Site	www.wuzr.com
Wattage	3,000
Format	Hot Country
News Director	Tom Lee	tomlee@originalcompany.com
Program Director	Jonathan Lange	jonathanlange@originalcompany.com
Notes	Serves Bicknell.

WWBL

Web Site	www.wwbl.com
Wattage	50,000
Format	Country
News Director	John Szink	johnszink@originalcompany.com
Program Director	Ken Booth	cowboyken@originalcompany.com
Notes	Serves Washington and has bureau there.

WYFX

Web Site	www.wyfxfm.com
Wattage	3,000
Format	Sports
Network	ESPN Radio
Program Director	Jonathan Lange	jonathanlange@originalcompany.com
Notes	Serves Mount Vernon & Evansville. Studio in Mount Vernon.

WZDM

Web Site	www.wzdm.com
Wattage	6,000
Format	Adult Contemporary
Program Director	Dave Young	daveyoung@originalcompany.com

See **WATI/American Family Radio** (listed under National Radio Stations).

WFML (96.7 FM)

Mailing Address	P.O. Box 574, Vincennes IN 47591
Street Address	1200 N. 2nd St., Vincennes IN 47591
Telephone	812-888-6829
Telephone (news)	812-888-5368
Fax	812-888-4955
E-mail	braddeetz@wfml.net
E-mail (news)	davefoster@wfml.net
Web Site	www.wfml.net
Wattage	3,000
Format	Variety Hits
Network Affiliations	Brownfield Farm Network, Dial-Global, Fox, Purdue University Basketball & Football
On-air Hours	24/7
Broadcast Company	DLC Media, Inc.
Owner/President	Dave Crooks	dlcmediainc@gmail.com
General Manager	Brad Deetz	braddeetz@wfml.net
Operations Manager	Andy Morrison	andy@wamwamfm.com
News Director	Dave Foster	davefoster@wfml.net
Sports Director	DeWayne Shake	dewayne@wamwamfm.com
Sales Manager	Beth Davis	beth@wamwamfm.com
Promotions Director	Katie Sullivan	katie@wamwamfm.com
Accepts PSAs?	yes (contact max@wfml.net)
Notes	Corporate offices in Washington, Indiana.

WVUB (91.1 FM)

Mailing Address	1002 N. 1st St., Vincennes IN 47591
Street Address	1200 N. 2nd St., Vincennes IN 47591
Telephone	812-888-5365
Telephone (news)	812-888-5368
Fax	812-882-2237
Fax (news)	812-888-5365
E-mail (news)	news@vinu.edu
Web Site	www.wvubhd.com
Wattage	50,000
Format	Hot Adult Contemporary/Contemporary Hit Radio
Network Affiliations	American Public Media, Network Indiana, NPR
On-air Hours	24/7
Owner	Board of Trustees for Vincennes University
General Mgr./Sales Mgr.	Phil Smith psmith@vinu.edu
News/Sports Director	Tony Cloyd tcloyd@vinu.edu
Program/Promotions Dir.	Michael Woods mwoods@vinu.edu
Accepts PSAs?	yes (contact Michael Woods)
Notes	Non-commercial station.

WVUT TV (Channel 22)

Address	1002 N. 2nd St., 64 Davis Hall, Vincennes IN 47591
Telephone	812-888-4345
Fax	812-882-2237
E-mail	wvut@vinu.edu
Web Site	www.vubroadcasting.org
Network Affiliation	PBS
On-air Hours	24/7
Owner	Vincennes University
General Manager	Al Rerko arerko@vinu.edu
Accepts PSAs?	yes (contact Al Rerko)
Notes	Non-commercial station.

WABASH

'the Paper' of Wabash County

Wabash County

Mailing Address	P.O. Box 603, Wabash IN 46992
Street Address	606 N. St. Rd. 13, Wabash IN 46992
Telephone	260-563-8326
Fax	260-563-2863
E-mail	ads@thepaperofwabash.com
E-mail (news)	news@thepaperofwabash.com
Web Site	www.thepaperofwabash.com
Publication Date	Weekly (Wednesday)
Circulation	16,200 (paid & free)
Publishing Company	The Paper Publishing
Publisher	Wayne Rees
General Manager	Michael Rees mrees@thepaperofwabash.com
Editor	Brent Swan news@thepaperofwabash.com

Wabash Plain Dealer

Mailing Address	P.O. Box 379, Wabash IN 46992
Street Address	123 W. Canal St., Wabash IN 46992
Telephone	260-563-2131
Telephone (toll-free)	800-659-6321
Fax	260-563-0816
E-mail (news)	news@wabashplaindealer.com
Web Site	www.wabashplaindealer.com
Publication Date	Daily (Monday-Saturday)
Circulation	5,400 (paid)
Publishing Company	Paxton Media Group
Publisher	Randy Mitchell
Editor	Joe Slacian
Advertising Manager	Karey Blakely
Sports Editor	Josh Sigler
TMC/Shopper	The Current (weekly)

WJOT (1510 AM) — WJOT (105.9 FM)

Address	1360 S. Wabash St., Wabash IN 46992
Telephone	260-563-1161
Telephone (toll-free)	866-563-1161
Fax	260-563-0883
E-mail	wjot@verizon.net
Web Site	www.1059thebash.com
Wattage	250 (WJOT AM)
	6,000 (WJOT FM)
Format	Classic Hits
Network Affiliations	ABC Oldies
On-air Hours	daytime (WJOT AM)
	24/7 (WJOT FM)
Broadcast Company	Mid-America Radio Group, Inc.
General Mgr./Sales Mgr.	Wade Weaver wade@1059thebash.com
News/Program Director	Andy McCord andy@realcountry1019.com
Accepts PSAs?	yes (contact Andy McCord)
Notes	WJOT AM & WJOT FM are simulcast.

WKUZ (95.9 FM)

Mailing Address	P.O. Box 342, Wabash IN 46992
Street Address	1864 S. Wabash St., STE A, Wabash IN 46992
Telephone	260-563-4111
Telephone (toll-free)	866-862-9590
Fax	260-563-4425
E-mail	info@959kissfm.net
Web Site	www.wkuz.com
Wattage	6,000
Format	Adult Contemporary
Network Affiliations	Jones
On-air Hours	24/7
Broadcast Company	Brothers Broadcasting, Inc.
General Mgr./Program Dir.	Dan McKay danmckayporat@gmail.com
Sports Director	Keith Martin keith@959kissfm.net
Accepts PSAs?	yes (contact Charlie Adams)
Notes	Does not broadcast local news.

WAKARUSA

Wakarusa Tribune
Elkhart County

Mailing Address	P.O. Box 507, Wakarusa IN 46573
Street Address	114 S. Elkhart St., Wakarusa IN 46573
Telephone/Fax	574-862-2179
E-mail	wakarusatribune@aol.com
Publication Date	Weekly (Wednesday)
Circulation	1,400 (paid)
Publisher	Mary Grantner
Editor	Bill Nich
Advertising Manager	Lee Garrett
Sports Editor	Eddie McDowell

WALKERTON

Walkerton Area Shopper
Saint Joseph County

Address	602 Roosevelt Rd., STE C, Walkerton IN 46574
Telephone/Fax	574-586-7467
E-mail	walkertonareashopper@embarqmail.com
Publication Date	Weekly (Monday)
Circulation	8,800 (free/delivered)
Publisher	Susan Rudecki

WARREN

Warren Weekly
Huntington County

Mailing Address	P.O. Box 695, Warren IN 46792
Telephone	260-375-3531
Telephone (toll-free)	877-811-9089
Fax	260-375-7007
E-mail	wwkly@citznet.com
Web Site	www.smalltownpapers.com
Publication Date	Weekly (Friday)
Circulation	3,200 (free/mailed)
Publisher/Editor	Nicki L. Zoda

WARSAW

'the Paper'—Bureau
Kosciusko County

Address	114 W. Market St., Warsaw IN 46580
Telephone	574-269-2932
Fax	574-269-5850
E-mail	warsaw@the-papers.com
Office Manager	Amii Bischof
Notes	Main office in Milford.

Times-Union

Mailing Address	P.O. Box 1448, Warsaw IN 46581
Street Address	Market & Indiana Streets, Warsaw IN 46580
Telephone	574-267-3111
Fax (news)	574-267-7784
Fax (advertising)	574-268-1300
Web Site	www.timesuniononline.com
Publication Date	Daily (Monday-Saturday)
Circulation	12,400 (paid)
Publishing Company	Reub Williams & Sons, Inc.
Publisher	Lane Williams Hartle
Editor	Gary Gerard	ggerard@timesuniononline.com
Advertising Manager	Bill Hays	bhays@timesuniononline.com
Sports Editor	Dale Hubler	dhubler@timesuniononline.com
Business Editor	Jen Gibson	jgibson@timesuniononline.com
TMC/Shopper	Extra (weekly)

WAWC (103.5 FM) — WRSW (1480 AM) — WRSW (107.3 FM)

Address	216 W. Market St., Warsaw IN 46580
Telephone	574-372-3064
Fax	574-267-2230
E-mail (news)	rgrossman@lakecityradio.com
On-air Hours	24/7
Broadcast Company	Talking Stick Communications
General Manager	Clint Marsh	cmarsh@lakecityradio.com
News/Sports Director	Roger Grossman	rgrossman@lakecityradio.com
Sales Manager	Dan Daggett	ddaggett@lakecityradio.com
Promotions Director	Chris Cage	ccage@lakecityradio.com
Accepts PSAs?	yes (contact Chris Cage)

WAWC

Web Site	www.willie1035.com
Wattage	3,000
Format	Country
Network	Fox News
Program Director	Jay Michaels	jmichaels@lakecityradio.com

WRSW AM

Web Site	www.1480sportsbug.net
Wattage	1,000
Format	Sports
Network	ESPN
Program Director	Roger Grossman	rgrossman@lakecityradio.com

WRSW FM

Web Site	www.wrsw.net
Wattage	50,000
Format	Classic Rock
Network	CNN
Program Director	Jay Michaels	jmichaels@lakecityradio.com

WIOE (98.3 FM)

Address	722 E. Center St., Warsaw IN 46580
Telephone	574-268-9830
E-mail	wioe@kconline.com
Web Site	www.wioe.com
Wattage	100
Format	Oldies
Network Affiliations	ABC
On-air Hours	24/7
Broadcast Company	Blessed Beginnings Broadcasting
General Mgr./Program Dir.	Brian Walsh
News Director	Brianna Walsh
Sports Director	Dave Baumgartner
Promotions Director	Jennifer Simpson
Accepts PSAs?	yes (contact Jennifer Simpson)
Notes	Non-commercial station.

See **WMYQ** (Fort Wayne).

WASHINGTON

Daviess County

Washington Times-Herald

Mailing Address	P.O. Box 471, Washington IN 47501	
Street Address	102 E. Van Trees St., Washington IN 47501	
Telephone	812-254-0480	
Telephone (toll-free)	800-235-4113	
Telephone (news)	812-254-0480	
Fax	812-254-7517	
E-mail (news)	patmorrison@washtimesherald.com	
Web Site	www.washtimesherald.com	
Publication Date	Daily (Monday-Saturday)	
Circulation	7,100 (paid)	
Publishing Company	Community Newspaper Holdings Inc.	
Publisher	Ron Smith	rsmith@washtimesherald.com
Managing Editor	Melody Brunson	mbrunson@washtimesherald.com
Advertising Manager	Stacey Ramsey	sramsey@washtimesherald.com
Sports Editor	Todd Lancaster	tlancaster@washtimesherald.com
News Editor	Pat Morrison	patmorrison@washtimesherald.com

WAMW (1580 AM) — WAMW (107.9 FM)

Address	800 W. National Hwy., Washington IN 47501
Telephone	812-254-6761
Telephone (toll-free)	877-254-9269
Fax	812-254-3940
E-mail	wamw@rtccom.net
E-mail (news)	taylor@wamwamfm.com
Web Site	www.wamwamfm.com
Network Affiliations	ABC Entertainment, Brownfield, Jones, Network Indiana
Broadcast Company	DLC Media, Inc.
President	Dave Crooks — dlcmediainc@gmail.com
General Manager	Brad Deetz — brad@wamwamfm.com
News Director	Taylor Brown — taylor@wamwamfm.com
Program Director	Andy Morrison — andy@wamwamfm.com
Sports Director	DeWayne Shake — dewayne@wamwamfm.com
Sales Manager	Beth Davis — beth@wamwamfm.com
Public Affairs Director	Lisa Jackman — lisa@wamwamfm.com
Promotions Director	Katie Sullivan — katie@wamwamfm.com
Accepts PSAs?	yes (contact Andy Morrison)

WAMW AM

Wattage	500
Format	Timeless
On-air Hours	daytime
Notes	Repeats on 95.9 FM (24/7)

WAMW FM

Wattage	3,000
Format	Classic Hits
On-air Hours	24/7

WBTO (102.3 FM) & WWBL (106.5 FM)—Bureau

Mailing Address	P.O. Box 616, Washington IN 47501
Street Address	3 E. Van Trees St., Washington IN 47501
Telephone	812-254-4300
Fax	812-254-4361
E-mail	wwbl@originalcompany.com
Notes	Main office in Vincennes.

WEST LAFAYETTE

See **Purdue Exponent** (listed under College Campus/Specialty Publications).

Tippecanoe County

WBAA (920 AM) — WBAA-FM (101.3 FM)

Address	712 Third St., West Lafayette IN 47907
Telephone	765-494-5920
Telephone (news)	765-494-3969
Fax	765-496-1542
E-mail	wbaa@wbaa.org
E-mail (news)	news@purdue.edu
Web Site	www.wbaa.org
Wattage	5,000 (WBAA AM)
	14,000 (WBAA-FM)
Format	News/Talk/Jazz/Information (WBAA AM)
	News/Classical Music (WBAA-FM)
Network Affiliations	NPR
On-air Hours	24/7
Owner	Purdue University
General Manager	Tim Singleton — tjsingle@purdue.edu
News/Sports/Public Affairs	Mike Loizzo — mloizzo@purdue.edu
Program Director	Greg Kostraba — gkostrab@purdue.edu
Music/Public Service Dir.	Jan Simon — cjsimon@purdue.edu
Accepts PSAs?	yes—only local PSAs (contact Jan Simon)
Notes	Non-commercial stations.

WLFI TV (Channel 18.1)

Address	2605 Yeager Rd., West Lafayette IN 47906
Telephone	765-463-1800
Telephone (toll-free)	800-877-9534
Fax	765-237-5001
Fax (news)	765-463-7979
Fax (sales)	765-497-2110
E-mail (news)	newsroom@wlfi.com
Web Site	www.wlfi.com
Network Affiliation	CBS
On-air Hours	24/7
Broadcast Company	LIN Television
General Manager	Tom Combs — tom.combs@wlfi.com
News Director	Chris Morisse — chris.morisse@wlfi.com
News Assignment Editor	Sue Scott — sue.scott@wlfi.com
Sports Director	Mike Cleff — mike.cleff@wlfi.com
General Sales Manager	Jenny Olszewski — jennyo@wlfi.com
Public Affairs/Promotions	Kurt Lahrman — kurt.lahrman@wlfi.com
Accepts PSAs?	yes (contact Kurth Lahrman)

WESTFIELD

See **Current in Westfield** (Carmel).

Hamilton County

WHITING

See **The WRite Stuff** (Valparaiso).

Lake County

WHITELAND

See **The Whiteland Times** (Greenwood).

WILKINSON

See **WRFM FM** (Greenfield).

WILLIAMSPORT

Review Republican

Mailing Address	P.O. Box 216, Williamsport IN 47993
Street Address	113 S. Perry St., Attica IN 47918
Telephone	765-762-3322
Telephone (toll-free)	800-301-4300
Fax	765-762-6418
E-mail	revrep@sbcglobal.net
Web Site	www.newsbug.info
Publication Date	Weekly (Thursday)
Circulation	2,400 (paid)
Publishing Company	Kankakee Valley Publishing Co.
Publisher	Don Hurd	dongo75@aol.com
Editor	Jane Jernagan	revrep@sbcglobal.net
Advertising Manager	Greg Willhite	atticasales@sbcglobal.net

WINAMAC

Pulaski County Express

Mailing Address	P.O. Box 218, Winamac IN 46996
Street Address	118 N. Market St., Winamac IN 46996
Telephone	574-946-7737
Telephone (toll-free)	888-946-2622
Fax	574-946-7763
E-mail (news)	kfritz@pulaskicountyexpress.com
Web Site	www.pulaskicountyexpress.com
Publication Date	Weekly (Saturday)
Circulation	8,000 (free/mailed)
Publishing Company	K & B Express Corp.
Publisher/Editor	Karen Fritz	kfritz@pulaskicountyexpress.com
Publisher/Advertising Mgr.	Brad Conn	ads@pulaskicountyexpress.com

Pulaski County Journal

Mailing Address	P.O. Box 19, Winamac IN 46996
Street Address	114 W. Main St., Winamac IN 46996
Telephone	574-946-6628
Fax	574-946-7471
E-mail	news@pulaskijournal.com
Web Site	www.pulaskijournal.com
Publication Date	Weekly (Wednesday)
Circulation	3,100 (paid)
Publishing Company	Pulaski County Press
Publisher	John Haley	haley@pulaskijournal.com
General Manager	Kari Stout	stout@pulaskijournal.com
Editor	Michelle Blevins	news@pulaskijournal.com
Advertising Manager	Tashia Lindvall	ads@pulaskijournal.com
TMC/Shopper	Independent (weekly)

See **WFRI** (Elkhart).

WINCHESTER

Randolph County

News-Gazette

Mailing Address	P.O. Box 429, Winchester IN 47394
Street Address	224 W. Franklin St., Winchester IN 47394
Telephone	765-584-4501
Telephone (toll-free)	800-782-2508
Fax	765-584-3066
E-mail	ngoffice@comcast.net
E-mail (news)	ngeditor@comcast.net
Web Site	www.winchesternewsgazette.com
Publication Date	Daily (Monday-Saturday)
Circulation	4,100 (paid)
Associate Publisher	Kami Shinn	ngadvertising@comcast.net
Managing Editor	Rick Reed	ngsports@comcast.net
City Editor	Bill Richmond	ngeditor@comcast.net
TMC/Shopper	Express (weekly)

See **WZZY** (Richmond).

WINFIELD

Lake County

Winfield American

Address	7590 E. 109th Ave., Winfield IN 46307
Telephone	219-662-8888
E-mail	mail@region-communications.com
Web Site	www.winfieldamerican.com
Publication Date	Weekly (Friday)
Circulation	5,000 (free/mailed)
Publishing Company	Region Communications
Publisher/Editor	Mike Gooldy
Advertising Manager	Mike Kucic

WOLCOTT

White County

New Wolcott Enterprise

Mailing Address	P.O. Box 78, Wolcott IN 47995
Street Address	125 W. Market St., Wolcott IN 47995
Telephone/Fax	219-279-2167
E-mail	wolcottenterprise@sugardog.com
Publication Date	Weekly (Thursday)
Circulation	800 (paid)
Owner/Publisher/Editor	Barbara Lawson
Advertising Manager	Deb Anker

ZIONSVILLE

Boone County

Zionsville Times-Sentinel

Address	250 S. Elm St., Zionsville IN 46077	
Telephone	317-873-6397	
Fax	317-873-6259	
E-mail (news)	news@timessentinel.com	
Web Site	www.timessentinel.com	
Publication Date	Weekly (Wednesday)	
Circulation	4,000 (paid)	
Publishing Company	cnhi media	
Publisher	Harold Allen	harold.allen@indianamediagroup.com
Managing Editor	Andrea Hirsch	andrea.hirsch@timessentinel.com
Advertising Manager	Bill Jarchow	bill.jarchow@indianamediagroup.com
Notes	Also publishes **Highflyer** (monthly advertising publication serving Hamilton County).	

AFRICAN-AMERICAN

See **Frost Illustrated** (Fort Wayne).

See **Gary Crusader Newspaper** (Gary).

See **Indiana Herald** (Indianapolis).

See **Indianapolis Recorder** (Indianapolis).

See **Ink Newspaper** (Fort Wayne).

See **Muncie Times** (Muncie).

See **Our Times** (Evansville).

AGRICULTURE

Farm World
Mailing Address	P.O. Box 90, Knightstown IN 46148
Street Address	27 N. Jefferson St., Knightstown IN 46148
Telephone	765-345-5133
Telephone (toll-free)	800-876-5133
Fax	800-318-1055
E-mail (news)	davidb@farmworldonline.com
Web Site	www.farmworldonline.com
Publication Date	Weekly (Wednesday)
Circulation	35,000 (paid)
Publishing Company	MidCountry Media, Inc.
Publisher	Tony Gregory — tony@antiqueweek.com
Editor/Associate Publisher	David Blower — davidb@farmworldonline.com
Ad Mgr./Assoc .Publisher	Toni Hodson — thodson@farmworldonline.com
Notes	Serves Illinois, Indiana, Kentucky, Michigan, Ohio and Tennessee.

The Farmer's Exchange
Mailing Address	P.O. Box 45, New Paris IN 46553
Street Address	19401 Industrial Dr., New Paris IN 46553
Telephone	574-831-2138
Fax	574-831-2131
E-mail (news)	jerry@farmers-exchange.net
Web Site	www.farmers-exchange.net
Publication Date	Weekly (Friday)
Circulation	13,100 (paid & free)
Publishing Company	Exchange Publishing Corp.
Publisher	Steve Yeater — steve@farmers-exchange.net
Editor	Jerry Goshert — jerry@farmers-exchange.net
Advertising Manager	Matt Yeater — matt@farmers-exchange.net

Hoosier Ag Today

Address	P.O. Box 34236, Indianapolis IN 46234
Telephone	317-247-9360
Fax	317-247-9380
President	Gary Truitt	gtruitt@hoosieragtoday.com
V. P. Operations	Andy Eubank	aeubank@hoosieragtoday.com
Chief Financial Officer	Kathleen Truitt	ktruitt@hoosieragtoday.com
Traffic Manager	Beth Carper	traffic@hoosieragtoday.com
Notes	Agriculture network serving Indiana radio stations.

Hoosier Farmer

Mailing Address	P.O. Box 1290, Indianapolis IN 46206
Street Address	225 S. East St., Indianapolis IN 46202
Telephone	317-692-7776
Telephone (toll-free)	800-327-6287
Telephone (news)	317-692-7824
Fax	317-692-7854
E-mail	askus@infarmbureau.org
E-mail (news)	kdutro@infarmbureau.org
Web Site	www.infarmbureau.org
Publication Date	Quarterly
Circulation	284,000 (paid)
Owner	Indiana Farm Bureau, Inc.
Publisher	Don Villwock
Editor	Andy Dietrick	adietrick@infarmbureau.org
Managing Editor	Kathleen Dutro	kdutro@infarmbureau.org

Indiana Agri-News

Address	2575 E. 55th Pl., STE A, Indianapolis IN 46220
Telephone	317-726-5391
Telephone (toll-free)	800-772-9354
Fax	317-726-5390
E-mail	editorial@agrinews-pubs.com
E-mail (advertising)	advertising@agrinews-pubs.com
Web Site	www.agrinews-pubs.com
Publication Date	Weekly (Friday)
Circulation	22,200 (paid)
Publishing Company	AgriNews Publications
Publisher	Lynn Barker	lbarker@agrinews-pubs.com
Editor	James Henry	jhenry@agrinews-pubs.com
Advertising Manager	Marguerite Allen	mallen@agrinews-pubs.com

Indiana Prairie Farmer

Mailing Address	P.O. Box 247, Franklin IN 46131
Telephone	317-738-0565
Fax	317-738-5441
E-mail	tbechman@farmprogress.com
Web Site	www.indianaprairiefarmer.com
Publication Date	Monthly
Circulation	29,500 (paid)
Publishing Company	Farm Progress Company
Editor	Tom Bechman	tbechman@farmprogress.com

Pets * Livestock

Mailing Address	P.O. Box 188, Milford IN 46542
Street Address	206 S. Main St., Milford IN 46542
Telephone	574-658-4111
Telephone (toll-free)	800-733-4111
Fax	574-658-4701
E-mail (news)	jseely@the-papers.com
Publication Date	Monthly
Circulation	7,000 (free)
Publishing Company	The Papers Inc.
Publisher	Ron Baumgartner — rbaumgartner@the-papers.com
Editor	Jeri Seely — jseely@the-papers.com
Associate Editor	Phoebe Muthart — pmuthart@the-papers.com
Advertising Manager	Vicky Howell — vhowell@the-papers.com
Notes	Serves Elkhart, Kosciusko & LaGrange counties.

ANTIQUES

AntiqueWeek

Mailing Address	P.O. Box 90, Knightstown IN 46148
Street Address	27 N. Jefferson St., Knightstown IN 46148
Telephone	765-345-5133
Telephone (toll-free)	800-876-5133
Fax	765-345-3398
Fax (advertising)	800-695-8153
E-mail (news)	connie@antiqueweek.com
Web Site	www.antiqueweek.com
Publication Date	Weekly (Monday)
Circulation	38,000 (paid)
Publishing Company	MidCountry Media, Inc.
Publisher	Tony Gregory — tony@antiqueweek.com
Managing Editor	Connie Swaim — connie@antiqueweek.com
Advertising Manager	Dan Morris — dmorris@antiqueweek.com
Notes	Serves Midwest states.

BUSINESS

Biz Voice

Mailing Address	P.O. Box 44926, Indianapolis IN 46244
Street Address	115 W. Washington St., STE 850 S, Indianapolis IN 46204
Telephone	317-264-3792
Fax	317-264-6855
E-mail	bizvoice@indianachamber.com
Web Site	www.bizvoicemagazine.com
Publication Date	Bi-monthly
Circulation	15,000 (free/mailed)
Owner	Indiana Chamber of Commerce
Publisher	Kevin Brinegar
Editor	Tom Schuman — tschuman@indianachamber.com
Managing Editor	Rebecca Patrick — rpatrick@indianachamber.com
Advertising Manager	Jim Wagner — jwagner@indianachamber.com

Business People

Address	7729 Westfield Dr., Fort Wayne IN 46825
Telephone	260-497-0433
Fax	260-497-0822
E-mail	jcopeland@businesspeople.com
E-mail (news)	editor@businesspeople.com
Web Site	www.businesspeople.com
Publication Date	Monthly
Circulation	9,000 (free/mailed)
Publishing Company	Michiana Business Publications
Publisher/Advertising Mgr.	Daniel C. Copeland dcopeland@businesspeople.com
Editor	Amber Recker arecker@businesspeople.com

Carmel Business Leader
Hendricks County Business Leader
Johnson County Business Leader

E-mail	info@businessleader.bz
Web Site	www.businessleader.bz
Publication Date	Monthly
Publisher	Rick Myers rick@businessleader.bz

Carmel Business Leader

Address	One S. Range Line Rd., STE 220, Carmel IN 46032
Telephone	317-489-4444
Fax	317-489-4446
Circulation	4,200 (free/mailed to every Carmel business)
Publishing Company	Current Publishing LLC

Hendricks County Business Leader

Address	2680 E. Main St., STE 219, Plainfield IN 46168
Telephone	317-837-5180
Fax	317-837-4901
Circulation	4,200 (free/mailed to every Hendricks County business)
Publishing Company	Times-Leader Publications LLC

Johnson County Business Leader

Address	301 Main St., Beech Grove IN 46107
Telephone	317-787-3291
Fax	317-787-3325
Circulation	4,900 (free/mailed to every Johnson County business)
Publishing Company	Times-Leader Publications LLC

Court and Commercial Record

Address	41 E. Washington St., STE 200, Indianapolis IN 46204
Telephone	317-636-0200
Fax	317-263-5259
E-mail (news)	rcollier@ibj.com
Web Site	www.courtcommercialrecord.com
Publication Date	Semi-weekly (Monday, Wednesday & Friday)
Circulation	1,300 (paid)
Publishing Company	IBJ Media
Publisher	Chris Katterjohn ckatterjohn@ibj.com
Editor	Rebecca Collier rcollier@ibj.com
Advertising Manager	Lisa Bradley lbradley@ibj.com

Evansville Business

Address	223 N.W. 2nd St., STE 200, Evansville IN 47708
Telephone	812-426-2115
Fax	812-426-2134
E-mail	jennifer@evansvilleliving.com
Web Site	www.evansvillebusiness.com
Publication Date	Bi-monthly
Circulation	10,000 (paid)
Publishing Company	Tucker Publishing Group
Publisher/Editor	Kristen Tucker	ktucker@evansvilleliving.com
Advertising Manager	Prudence Hoesli	prudence@evansvilleliving.com

Fueling Indiana — Propane Express

Address	115 W. Washington St., STE 1690S, Indianapolis IN 46204
Telephone (advertising)	502-423-7272
Fax	317-630-1827
Fax (advertising)	502-423-7979
Publication Date	Quarterly
Publishing Company	Innovative Publishing Ink

Fueling Indiana

Telephone	317-633-4662
Web Site	www.ipca.org
Circulation	2,000 (paid)
Editor	Juva Sizemore Barber	jbarber@ipca.org
Advertising Manager	Kelly Arvin	karvin@ipipublishing.com
Notes	Covers petroleum industry in Indiana.

Propane Express

Telephone	317-655-4444
Telephone (toll-free)	877-277-6726
Web Site	www.indianapropane.com
Circulation	1,000 (paid)
Editor	Juva Sizemore Barber	jbarber @indianapropane.com
Advertising Manager	Jerry Stains	jstains@ipipublishing.com
Notes	Covers propane industry in Indiana.

Greater Fort Wayne Business Weekly

Mailing Address	P.O. Box 11448, Fort Wayne IN 46858
Street Address	826 Ewing St., Fort Wayne IN 46802
Telephone	260-426-2640
Fax	260-426-2503
E-mail (news)	news@fwbusiness.com
Web Site	www.fwbusiness.com
Publication Date	Weekly (Friday)
Circulation	5,000 (paid)
Publishing Company	KPC Media Group
Publisher	Terry Housholder	terryh@kpcnews.net
General Manager	Lynn Sroufe	lsroufe@kpcnews.net
Editor	Barry Rochford	barryr@kpcnews.net

Hamilton County Business Magazine

Mailing Address	P.O. Box 502, Noblesville IN 46061
Street Address	802 Mulberry St., STE BB, Noblesville IN 46060
Telephone/Fax	317-774-7747
E-mail (news)	news@hamiltoncountybusiness.com
Web Site	www.hamiltoncountybusiness.com
Publication Date	Bi-monthly
Circulation	3,300 (free & paid)
Publishing Company	Hamilton County Media Group
Publisher/Editor	Michael Corbett mcorbett@hamiltoncountybusiness.com
Notes	Also publishes **Welcome to Hamilton County Visitors Guide** (annual publication).

Hoosier Banker

Address	6925 Parkdale Pl., Indianapolis IN 46254
Telephone	317-387-9380
Fax	317-387-9374
E-mail	Lwilson@indianabankers.org
Web Site	www.indianabankers.org
Publication Date	Monthly
Circulation	4,700 (free to members)
Owner	Indiana Bankers Association
Publisher	S. Joe DeHaven jdehaven@indianabankers.org
Editor	Laura Wilson lwilson@indianabankers.org

Indiana Beverage Journal

Mailing Address	P.O. Box 5067, Zionsville IN 46077
Street Address	7379 Fox Hollow Ridge, Zionsville IN 46077
Telephone	317-733-0527
Fax	317-733-0220
E-mail	ibjzstew@indy.rr.com
Web Site	www.bevnetwork.com
Publication Date	Monthly
Circulation	3,200 (paid)
Publishing Company	Indiana Beverage Life, Inc.
Publisher	Stewart Baxter
Notes	Covers alcoholic beverage industry in Indiana.

Indiana Minority Business Magazine

Address	2901 N. Tacoma Ave., Indianapolis IN 46218
Telephone	317-924-5143
Fax	317-921-6653
Fax (news)	317-924-5148
E-mail	hiphopchatter@yahoo.com
Web Site	www.indianaminoritybusinessmagazine.com
Publication Date	Quarterly
Circulation	55,000 (paid & free)
Vice President	Rickie Clark hiphopchatter@yahoo.com

Indiana REALTOR

Address	7301 N. Shadeland Ave., STE A, Indianapolis IN 46250
Telephone	317-842-0890
Telephone (toll-free)	800-284-0084
E-mail	sahartman@indianarealtors.com
Web Site	www.indianarealtors.com
Publication Date	Quarterly
Circulation	18,000 (free/mailed)
Managing Editor/Ad Mgr.	Stacey Hartman sahartman@indianarealtors.com

Indianapolis Business Journal

Address	41 E. Washington St., STE 200, Indianapolis IN 46204	
Telephone	317-634-6200	
Telephone (toll-free)	800-968-1225	
Fax	317-263-5060	
Fax (news)	317-263-5406	
Fax (advertising)	317-263-5400	
E-mail	info-ibj@ibj.com	
Web Site	www.ibj.com	
Publication Date	Weekly (Monday)	
Circulation	16,300 (paid)	
Publishing Company	IBJ Media	
Publisher	Chris Katterjohn	ckatterjohn@ibj.com
Editor	Tom Harton	tharton@ibj.com
Managing Editor	Greg Andrews	gandrews@ibj.com
Advertising Manager	Lisa Bradley	lbradley@ibj.com

Inside Indiana Business with Gerry Dick

Address	1630 N. Meridian St., STE 400, Indianapolis IN 46202
Telephone	317-275-2010
E-mail	newsletter@growindiana.net
Web Site	www.insideindianabusiness.com
Notes	News service covering Indiana business.

Tribune Business Weekly

Address	225 W. Colfax Ave., South Bend IN 46626	
Telephone	574-235-6302	
Fax	574-239-2646	
E-mail	biznews@sbtinfo.com	
Web Site	www.southbendtribune.com	
Publication Date	Weekly (Monday)	
Circulation	9,000 (paid)	
Publishing Company	Schurz Communications	
Publisher	David Ray	dray@sbtinfo.com
Editor	Ed Semmler	esemmler@sbtinfo.com
TBW Coordinator	Joni Gibley	jgibley@sbtinfo.com
Advertising Manager	Kim Keigley	kkeigley@sbtinfo.com

Wabash Valley Journal of Business

Address	2120 Second Ave., Terre Haute IN 47807
Telephone	317-641-7759
Telephone (advertising)	812-223-2522
E-mail	thjournal@thjournal.com
E-mail (news)	robertf@thjournal.com
Web Site	www.thjournal.com
Publication Date	Monthly
Circulation	9,000 (free)
Publishing Company	Flottsom Communications
Publisher/Editor	Robert Flott robertf@thjournal.com
Advertising Manager	John Gifford johng@thjournal.com
Notes	Also publishes **Wabash Valley Christian Business Directory** (annual publication).

COLLEGE ALUMNI

Indiana Alumni Magazine

Address	1000 E. 17th St., Bloomington IN 47408
Telephone	812-855-5785
Telephone (toll-free)	800-824-3044
Fax	812-855-4228
E-mail	iualumni@indiana.edu
Web Site	www.alumni.indiana.edu/magazine/
Publication Date	Bi-monthly
Circulation	77,700 (paid with membership)
Owner	Indiana University Alumni Assn.
Publisher	Tom Martz
Editor	Mike Wright miwright@indiana.edu
Managing Editor	J. D. Denny josdenny@indiana.edu

Purdue Alumnus

Address	403 W. Wood St., West Lafayette IN 47907
Telephone	765-494-5175
Telephone (toll-free)	800-414-1541
Fax	765-494-9179
Fax (news & advertising)	765-494-8290
E-mail	alumnus@purdue.edu
Web Site	www.purduealum.org
Publication Date	Bi-monthly
Circulation	60,000 (paid)
Owner	Purdue Alumni Association
Publisher	Kirk Cerny
Editor	Kelly Hiller khiller@purdue.edu
Managing Editor	Nicki Reas nreas@purdue.edu
Advertising Manager	Dan Rhodes darhodes@purdue.edu

COLLEGE CAMPUS

Bachelor

Address	301 W. Wabash Ave., Crawfordsville IN 47933
Telephone	765-361-6213
Web Site	www.bachelor.wabash.edu
Publication Date	Weekly (Friday)
Circulation	1,200 (free)
Owner	Wabash College
Faculty Advisor	Howard Hewitt hewitth@wabash.edu

Ball State Daily News

Address	AJ 276, Ball State University, Muncie IN 47306
Telephone	765-285-8256
Telephone (news)	765-285-8255
Fax	765-285-8248
E-mail (news)	news@bsudailynews.com
E-mail (advertising)	dailynewsads@bsu.edu
E-mail (sports)	sports@bsudailynews.com
Web Site	www.bsudailynews.com
Publication Date	Daily (Monday-Thursday)
Circulation	10,000-free (August-May); 5,000-free (summer)
Owner	Ball State University
Faculty Advisor	John Strauss jcstrauss@bsu.edu

Butler Collegian

Address	4600 Sunset Ave., Fairbanks RM 210, Indianapolis IN 46208
Telephone	317-940-8813
Telephone (advertising)	317-940-9358
Fax	317-940-9713
E-mail	collegian@butler.edu
Publication Date	Weekly (Wednesday)
Circulation	3,000 (free/newsstand)
Owner	Butler University
Faculty Advisor	Charles St. Cyr cstcyr@butler.edu

Chronicle

Address	2200 169th St., SUL 344-H, Hammond IN 46323
Telephone	219-989-2547
E-mail	chronicle_eic@yahoo.com
Web Site	www.pucchronicle.com
Publication Date	Weekly (Monday)
Circulation	2,000 (free/newsstand)
Owner	Purdue University Calumet

Communicator

Address	2101 E. Coliseum Blvd., Walb Union STE 215, Fort Wayne IN 46805
Telephone (news)	260-487-6584
Fax	260-481-6045
E-mail	contact@ipfwcommunicator.org
E-mail (advertising)	ads@ipfwcommunicator.org
Web Site	www.ipfwcommunicator.org
Publication Date	Weekly-Wednesday (during school year)
Circulation	4,000 (free)
Owner	Indiana University-Purdue University Fort Wayne
Publisher	Matt McClure publisher@ipfwcommunicator.org

Crescent Magazine

Address	1800 Lincoln Ave., Evansville IN 47722
Telephone	812-488-2846
Fax	812-488-2224
E-mail	crescentmagazine@evansville.edu
Web Site	www.uecrescentmagazine.com
Publication Date	monthly
Circulation	1,800 (free/newsstand)
Owner	University of Evansville

The DePauw

Address	609 S. Locust St., Greencastle IN 46135
Telephone	765-658-5972
E-mail	editor@thedepauw.com
Web Site	www.thedepauw.com
Publication Date	Semi-weekly (Tuesday & Friday)
Circulation	1,800 (free/newsstand)
Publishing Company	The DePauw
Notes	Serves DePauw University.

Earlham Word

Mailing Address	Earlham College, Drawer #273, Richmond IN 47374
Street Address	Earlham College, 801 National Rd. W., Richmond IN 47374
Telephone	765-983-1569
E-mail	word-l@earlham.edu
Web Site	www.ecword.org
Publication Date	Weekly (Friday)
Circulation	1,300 (free)
Owner	Earlham College
Faculty Advisor	Judi Hetrick hetriju@earlham.edu

Franklin Newspaper

Address	Pulliam School of Journalism, 101 Branigin Blvd., Franklin IN 46131
Telephone	317-738-8191
E-mail	thefranklin@franklincollege.edu
Web Site	www.thefranklinonline.com
Publication Date	Weekly (Friday)
Circulation	1,000 (free/delivered)
Owner	Franklin College
Faculty Advisors	John Krull jkrull@franklincollege.edu
	Hank Nuwer
	Dennis Cripe

Goshen College Record

Address	1700 S. Main St., Goshen IN 46526
Telephone	574-535-7398
Fax	574-535-7293
E-mail	record@goshen.edu
Web Site	www.goshen.edu/record
Publication Date	Weekly: Thursday (during school year)
Circulation	1,000 (free/newsstand)
Owner	Goshen College
Faculty Advisor	Duane Stoltzfus dstoltzfus@goshen.edu

Indiana Daily Student

Address	940 E. 7th St., 120 Ernie Pyle Hall, Bloomington IN 47405
Telephone	812-855-0763
Telephone (news)	812-855-0760
Fax	812-855-8009
E-mail	ids@indiana.edu
Web Site	www.idsnews.com
Publication Date	Daily (Monday-Friday)
Circulation	15,500 (free/delivered & newsstand)
Owner	Indiana University
Publisher	Ron Johnson ronejohn@indiana.edu
Advertising Manager	Amy Swain ads@idsnews.com

Indiana Statesman

Address	HMSU RM 716, Indiana State University, Terre Haute IN 47809
Telephone	812-237-3025
Telephone (news)	812-237-3289
Fax	812-237-7629
E-mail	saseditr@isugw.indstate.edu
Web Site	www.indianastatesman.com
Publication Date	Semi-weekly (Monday, Wednesday & Friday)
Circulation	5,000 (free/newsstand)
Publishing Company	Indiana State University Student Publications
General Manager	Merv Hendricks mhendricks2@isugw.indstate.edu
Assistant Director, Student Publications	Marcy Shonk mshonk1@isugw.indstate.edu

IUPUI Student Media

Address	535 W. Michigan St., IT 557, Indianapolis IN 46202
Telephone	317-278-5332
E-mail	iupuistudentmedia@gmail.com
Web Site	www.iupuistudentmedia.com
Publication Date	Quarterly
Circulation	7,500
Owner	IUPUI School of Journalism
Publisher	Maggie Balough Hillery mbalough@iupui.edu
Advertising Manager	Blake Egan bsegan@iupui.edu

Observer

Mailing Address	P.O. Box 779, Notre Dame IN 46556
Street Address	024 South Dining Hall, U. of Notre Dame, Notre Dame IN 46556
Telephone	574-631-7471
Telephone (news)	574-631-5323
Telephone (advertising)	574-631-6900
Fax	574-631-6927
E-mail	obsnews@nd.edu
Web Site	www.ndsmcobserver.com
Publication Date	Daily (Monday-Friday)
Circulation	10,000
Owner	University of Notre Dame
Notes	Serves University of Notre Dame and St. Mary's College.

Purdue Exponent

Mailing Address	P.O. Box 2506, West Lafayette IN 47996
Street Address	460 Northwestern Ave., West Lafayette IN 47906
Telephone	765-743-1111
Fax	765-743-6087
E-mail	help@purdueexponent.org
E-mail (news)	editor@purdueexponent.org
Web Site	www.purdueexponent.org
Publication Date	Daily (Monday-Friday)
Circulation	18,000 (free/delivered & newsstand)
Publishing Company	Purdue Student Publishing Foundation

The Reflector

Address	1400 E. Hanna Ave., 333 Esch Hall, Indianapolis IN 46227
Telephone	317-788-3269
Fax	317-788-3490
E-mail	reflector@uindy.edu
Web Site	www.reflector.uindy.edu
Publication Date	Bi-weekly-Wednesday (during school year)
Circulation	14,00 (free/newsstand)
Owner	University of Indianapolis
Faculty Advisor	Jeanne Criswell jcriswell@uindy.edu

The Shield

Address	8600 University Blvd., Evansville IN 47712
Telephone (news)	812-465-1645
Telephone (advertising)	812-464-1870
Fax	812-465-1632
E-mail	shield@usi.edu
Web Site	www.usishield.com
Publication Date	Weekly (Thursday)
Circulation	2,500 (free/newsstand)
Owner	University of Southern Indiana
Faculty Advisor	Erin Gibson emgibson@usi.edu

Torch

Address	35 Schnabel, 1809 Chapel Dr., Valparaiso IN 46383
Telephone	219-464-5426
Fax	219-464-6728
E-mail	torch@valpo.edu
E-mail (news)	torch.news@valpo.edu
E-mail (advertising)	torch.advertising@valpo.edu
E-mail (sports)	torch.sports@valpo.edu
Web Site	www.valpo.edu/torch
Publication Date	Weekly (Friday)
Circulation	3,500 (free/newsstand)
Owner	Valparaiso University

CONSTRUCTION

Indiana Builder News

Address	101 W. Ohio St., STE 1111, Indianapolis IN 46204
Telephone	317-917-1100
Telephone (toll-free)	800-377-6334
Fax	317-917-0335
E-mail	info@buildindiana.org
E-mail (news)	cindy@buildindiana.org
Web Site	www.buildindiana.org
Publication Date	Bi-monthly
Circulation	5,500 (paid with membership)
Owner	Indiana Builders Association
Publisher/Editor/Ad Mgr.	Cindy Bussell cindy@buildindiana.org

Indiana Contractor

Address	9595 Whitley Dr., STE 208, Indianapolis IN 46240
Telephone	317-575-9292
Fax	317-575-9378
E-mail	brenda@iaphcc.com
Web Site	www.iaphcc.com
Publication Date	Quarterly
Circulation	5,000 (free/delivered)
Owner	Indiana Association of Plumbing-Heating-Cooling Contractors
Publisher/Editor	Brenda Dant brenda@iaphcc.com

DINING

Indianapolis Dine

Address	120 E. Vermont St., Indianapolis IN 46204
Telephone	317-333-7200
Fax	317-333-7207
Web Site	www.indianapolisdine.com
Publication Date	Bi-monthly
Circulation	38,000 (paid/mailed & newsstand)
Editor	Courtney Leach cleach@indianapolisdine.com

DIVERSITY

Indiana Diversity Focus Newspaper

Address	6539 Camarillo Ct., Indianapolis IN 46278
Telephone	317-490-4950
Fax	317-293-3989
E-mail	hiphopchatter@yahoo.com
Web Site	www.diversitychangeresults.com
Publication Date	Monthly
Circulation	44,100 (paid & free)
President/CEO	Rickie Clark hiphopchatter@yahoo.com

EDUCATION

Indiana Education Insight

Mailing Address	P.O. Box 383, Noblesville IN 46061
Telephone	317-955-9997
Fax	317-955-9998
E-mail	adam.vanosdol@ingrouponline.com
Web Site	www.ingrouponline.com
Publication Date	Bi-weekly
Circulation	paid
Publishing Company	INGroup
Publisher	Edward D. Feigenbaum EDF@ingrouponline.com
Editor	Adam Van Osdol adam.vanosdol@ingrouponline.com
Notes	Advertising accepted for website only.

ISTA Advocate

Address	150 W. Market St., STE 900, Indianapolis IN 46204
Telephone	317-263-3400
Telephone (toll-free)	800-382-4037
Fax	317-655-3700
Web Site	www.ista-in.org
Publication Date	Quarterly
Circulation	54,000 (paid with membership)
Owner	Indiana State Teachers Assn.
Publisher/Executive Editor	Mark Shoup mshoup@ista-in.org
Editor/Advertising Mgr.	Kathleen Berry kberry@ista-in.org

ENVIRONMENTAL

Indiana Living Green

Address	1730 S. 950 E., Zionsville IN 46077
Telephone	317-769-3456
E-mail	info@IndianaLivingGreen.com
E-mail (news)	editor@IndianaLivingGreen.com
Web Site	www.IndianaLivingGreen.com
Publication Date	Bi-monthly
Circulation	25,000 (free/newsstand)
Publisher	Lynn Jenkins lynn@IndianaLivingGreen.com
Editor	Jo Ellen Meyers Sharp editor@IndianaLivingGreen.com

GOVERNMENT

Indiana Gaming Insight — Indiana Legislative Insight

Mailing Address	P.O. Box 383, Noblesville IN 46061
Telephone	317-817-9997
Fax	317-817-9998
E-mail	info@ingrouponline.com
Web Site	www.ingrouponline.com
Publication Date	Bi-weekly (Indiana Gaming Insight)
	Weekly (Indiana Legislative Insight)
Circulation	paid
Publishing Company	INGroup
Publisher/Editor	Edward D. Feigenbaum EDF@ingrouponline.com

HEALTH CARE

Indiana Pharmacist

Address	729 N. Pennsylvania St., Indianapolis IN 46204
Telephone	317-634-4968
Fax	317-632-1219
E-mail	tabitha@indianapharmacists.org
Web Site	www.indianapharmacists.org
Publication Date	Quarterly
Circulation	1,500 (paid)
Owner	Indiana Pharmacists Alliance
Editor	Lawrence Sage
Managing Editor	Tabitha Cross tabitha@indianapharmacists.org

HISPANIC

See **El Mexicano Newspaper** (Fort Wayne).

See **El Puente** (Goshen).

See **en Espanol** (listed with Journal and Courier, Lafayette).

See **Info: Logansport's Bilingual Newspaper** (Logansport).

See **La Ola Latino-Americana** (Indianapolis).

See **La Voz de Indiana** (Indianapolis).

HISTORIC PRESERVATION

Indiana Preservationist

Address	340 W. Michigan St., Indianapolis IN 46202
Telephone	317-639-4534
Telephone (toll-free)	800-450-4534
Fax	317-639-6734
E-mail	info@historiclandmarks.org
Web Site	www.historiclandmarks.org
Publication Date	Bi-monthly
Circulation	8,800 (paid with membership)
Owner	Historic Landmarks Foundation of Indiana
Staff Writer	Paige Wassel editor@historiclandmarks.org

LABOR

Indiana Labor News
Address	2620 E. 10th St., Indianapolis IN 46201
Telephone	317-264-4275
Telephone (toll-free)	800-428-8842
Fax	317-264-4280
E-mail (news)	labornewspaper@sbcglobal.net
E-mail (advertising)	labornewsads@yahoo.com
Web Site	www.labornews.com
Publication Date	Monthly
Publisher	Fred Levin
Editor	Sandra Robinson

LEGAL

Indiana Lawyer
Address	41 E. Washington St., STE 200, Indianapolis IN 46204	
Telephone	317-636-0200	
Telephone (toll-free)	800-968-1225	
Fax	317-263-5259	
Fax (advertising)	317-263-5400	
E-mail (news)	rcollier@ibj.com	
Web Site	www.theindianalawyer.com	
Publication Date	Bi-weekly (Wednesday)	
Circulation	6,700 (paid)	
Publishing Company	IBJ Media	
Publisher/Editor	Rebecca Collier	rcollier@ibj.com
Advertising Manager	Lisa Bradley	lbradley@ibj.com

NATURAL LIVING

Branches Magazine
Mailing Address	P.O. Box 30920, Indianapolis IN 46230	
Telephone	317-255-5594	
E-mail (news)	editor@branches.com	
Web Site	www.branches.com	
Publication Date	Bi-monthly	
Circulation	20,000 (free/newsstand)	
Publishing Company	Apple Press, Inc.	
Publisher/Advertising Mgr.	Thomas P. Healy	apple@branches.com
Editor	Elsa Kramer	editor@branches.com

PARENTING

Columbus Parent
Address	333 Second St., Columbus IN 47201	
Telephone	812-379-5625	
Fax	812-379-5776	
E-mail	dshowalter@therepublic.com	
Web Site	www.therepublic.com	
Publication Date	Bi-monthly	
Circulation	10,000 (free)	
Publishing Company	Home News Enterprises	
Editor	Doug Showalter	dshowalter@therepublic.com

Family Life

Address	2120 Second Ave., Terre Haute IN 47807
Telephone/Fax	317-641-7759
Telephone (advertising)	812-223-2522
Web Site	www.wvfamilylife.com
Publication Date	Monthly
Circulation	5,000 (free)
Publishing Company	Flottsom Communications
Publisher	Robert Flott	robertf@thjournal.com
Editor	Jonathan Moore	jonathanm@thjournal.com
Advertising Manager	John Gifford	johng@thjournal.com

Greater Fort Wayne Family

Address	826 Ewing St., Fort Wayne IN 46802
Telephone	260-426-2640
Fax	260-426-2503
E-mail	graceh@fwfamily.com
Web Site	www.fwfamily.com
Publication Date	Monthly
Circulation	16,500 (free)
Publishing Company	KPC Media Group
Publisher	Terry Housholder	terryh@kpcnews.net
General Manager	Lynn Sroufe	lsroufe@kpcnews.net
Editor	Grace Housholder	graceh@fwfamily.com
Advertising Manager	Sherri Ayres	sayres@kpcnews.net

Indiana Parenting Magazine

Address	6539 Camarillo Ct., Indianapolis IN 46278
Telephone	317-490-4950
Fax	317-293-3989
E-mail	hiphopchatter@yahoo.com
Web Site	www.indianaparenting.com
Publication Date	Bi-monthly
Circulation	34,000 (paid & free)
General Manager	Jennifer Banks
Vice President	Rickie Clark	hiphopchatter@yahoo.com

Indy's Child Parenting Magazine

Address	921 E. 86th St., STE 130, Indianapolis IN 46240
Telephone	317-722-8500
Fax	317-722-8510
E-mail	indyschild@indyschild.com
Web Site	www.indyschild.com
Publication Date	Monthly
Circulation	110,000 (free/newsstand)
Publishing Company	Midwest Parenting Publications
Publisher	Mary Wynne-Cox	mary@indyschild.com
Executive Editor	Lynette Rowland	editor@indyschild.com

PETS

Indy Tails Pet Magazine

Address	4527 N. Ravenswood Ave., Chicago IL 60640
Telephone	317-538-7115
Telephone (toll-free)	866-803-8245
Fax	773-561-3030
E-mail	JB@TailsInc.com
Web Site	www.TailsInc.com
Publication Date	Monthly
Circulation	35,000 (free/newsstand)
Publishing Company	Tails Pet Media Group, Inc.
Publisher	Al Brown	al@TailsInc.com
Editor	Janice Brown	JB@TailsInc.com
Managing Editor	Lauren Lewis	lauren@TailsInc.com
Advertising Manager	Erin Knauss	erin@TailsInc.com

RECREATION

Gad-A-Bout

Address	105 E. South St., Centerville IN 47330
Telephone/Fax	765-855-3857
Telephone (toll-free)	877-855-4237
E-mail	thegadabout@verizon.net
Publication Date	Monthly
Circulation	10,000 (free/newsstand)
Publisher	Ray Dickerson
Notes	Focuses on the outdoors in Indiana and surrounding states.

Indiana Outdoor News

Mailing Address	P.O. Box 69, Granger IN 46530
Street Address	16828 Barry Knoll Way, Granger IN 46530
Telephone	574-273-5160
Telephone (toll-free)	877-251-2112
Fax	800-496-8075
E-mail	contact@raghorn.com
E-mail (news)	submit@raghorn.com
Web Site	www.IndianaOutdoorNews.net
Publication Date	Monthly
Circulation	75,000 (free & paid/mailed & newsstand)
Publishing Company	Raghorn Inc.
Publisher/Advertising Mgr.	Brian Smith	publisher@raghorn.com
Editor	Joshua Lantz	editor1@raghorn.com

Tri-State Outdoor News

Mailing Address	P.O. Box 30, Princeton IN 47670
Street Address	100 N. Gibson St., Princeton IN 47670
Telephone	812-385-2525
Fax	812-386-6199
E-mail	outdoornews@pdclarion.com
Web Site	www.tristate-media.com
Publication Date	Monthly
Circulation	12,000 (free)
Publishing Company	Princeton Publishing
Publisher	Gary Blackburn	gblack@pdclarion.com
Editor/Advertising Mgr.	Mark Crowley	outdoornews@pdclarion.com

RELIGION

The Catholic Moment

Mailing Address	P.O. Box 1603, Lafayette IN 47902
Street Address	610 Lingle Ave., Lafayette IN 47901
Telephone	765-742-2050
Telephone (toll-free)	800-942-2397
Fax	765-742-7513
E-mail	moment@dioceseoflafayette.org
Web Site	www.thecatholicmoment.org
Publication Date	Weekly (Sunday)
Circulation	28,000 (paid/mailed)
Owner	Diocese of Lafayette-in-Indiana
Publisher	Most Rev. William L. Higi
Editor	Kevin Cullen	kcullen@dioceseoflafayette.org
Managing Editor	Laurie Cullen	lcullen@dioceseoflafayette.org
Advertising/Circulation Mgr.	Carolyn McKinney	cmckinne@dioceseoflafayette.org
Contributing Editor	Caroline Mooney	cmooney@dioceseoflafayette.org
Notes	Serves Roman Catholics in North Central Indiana.

Criterion

Mailing Address	P.O. Box 1717, Indianapolis IN 46206
Street Address	1400 N. Meridian St., Indianapolis IN 46202
Telephone	317-236-1570
Telephone (toll-free)	800-382-9836
Telephone (news)	317-236-1585
Fax	317-236-1593
Fax (advertising)	317-236-1434
E-mail	criterion@archindy.org
Web Site	www.criteriononline.com
Publication Date	Weekly (Friday)
Circulation	72,000 (paid)
Owner	Roman Catholic Archdiocese of Indianapolis
Publisher	Archbishop Daniel M. Buechlein
Managing Editor	Mike Krokos	mkrokos@archindy.org
Advertising Manager	Ron Massey	rmassey@archindy.org
Notes	Serves Roman Catholics of Archdiocese of Indianapolis.

Indiana Jewish Post & Opinion

Address	238 S. Meridian St., STE 502, Indianapolis IN 46225
Telephone	317-972-7800
Fax	317-972-7807
E-mail	jpostopinion@gmail.com
Web Site	www.jewishpostopinion.com
Publication Date	Semi-monthly (Wednesday)
Circulation	14,000 (paid)
Owner/Publisher	Jennie Cohen	jpostopinion@gmail.com
Advertising Manager	Barbara Lemaster	ads@indy.rr.com

Message

Mailing Address	P.O. Box 4169, Evansville IN 47724
Street Address	4200 N. Kentucky Ave., Evansville IN 47711
Telephone	812-424-5536
Telephone (toll-free)	800-637-1731
Fax	812-424-0972
E-mail	message@evansville-diocese.org
Web Site	www.evansville-diocese.org
Publication Date	Weekly (Friday)
Circulation	5,200 (paid)
Owner	Catholic Press of Evansville
Publisher	Most Rev. Gerald Gettelfinger
Editor	Paul R. Leingang
Advertising Representative	Carol Funke
Notes	Serves Catholics in 12 southwestern Indiana counties.

Northwest Indiana Catholic

Address	9292 Broadway, Merrillville IN 46410
Telephone	219-769-9292
Fax	219-738-9034
E-mail	nwic@dcgary.org
Web Site	www.nwicatholic.com
Publication Date	Weekly
Circulation	16,000 (paid)
Owner	Catholic Diocese of Gary
Publisher	Bishop Dale J. Melczek
General Manager/Editor	Steve Euvino seuvino@dcgary.org
Notes	Serves Roman Catholics in Lake, LaPorte, Porter and Starke counties.

Today's Catholic

Mailing Address	P.O. Box 11169, Fort Wayne IN 46856
Street Address	915 S. Clinton St., Fort Wayne IN 46802
Telephone	260-456-2824
Fax	260-744-1473
E-mail (news)	editor@fw.diocesefwsb.org
Web Site	www.todayscatholicnews.org
Publication Date	Weekly (Wednesday)
Circulation	53,000 (free to members/paid for others)
Owner	Catholic Diocese of Fort Wayne-South Bend
Publisher	Bishop John D'Arcy
Editor	Tim Johnson editor@fw.diocesefwsb.org
Advertising Manager	Kathy Denice kdenice@diocesefwsb.org
Notes	Serves Catholic Diocese of Fort Wayne-South Bend. Bureau: 114 W. Wayne St., South Bend IN 46601; 574-234-0687

SENIOR CITIZENS

Over Fifty Magazine
Mailing Address	P.O. Box 4128, Lawrenceburg IN 47025
Street Address	126 W. High St., Lawrenceburg IN 47025
Telephone	812-537-0063
Fax	812-537-5576
Publication Date	Monthly
Circulation	7,000 (free/mailed & newsstand)
Publishing Company	Register Publications
Publisher	Joe Awad	editor@registerpublications.com
Editor	Robin Duke	fiftiespubs@registerpublications.com
Advertising Manager	Loretta Day	lday@registerpublications.com

Senior Life
Mailing Address	P.O. Box 188, Milford IN 46542
Street Address	206 S. Main St., Milford IN 46542
Telephone	574-658-4111
Telephone (toll-free)	800-733-4111
Fax	574-658-4701
E-mail (news)	jseely@the-papers.com
Web Site	www.the-papers.com
Publication Date	Monthly
Circulation	24,600-free/newsstand (Allen County edition)
	16,700-free/newsstand (Elkhart-Kosciusko County edition)
	33,600-free/newsstand (Indianapolis edition)
	26,500-free/newsstand (Northwest-Merrillville & Valparaiso edition)
	25,000-free/newsstand (Saint Joseph County edition)
Publishing Company	The Papers Inc.
Publisher	Ron Baumgartner	rbaumgartner@the-papers.com
General Manager	Collette Knepp	cknepp@the-papers.com
Editor	Jeri Seely	jseely@the-papers.com
Associate Editor	Deb Patterson	dpatterson@the-papers.com
Advertising Manager	Vicky Howell	vhowell@the-papers.com
Ad Mgr. (Indianapolis)	Kim Gross	kgross@the-papers.com
Assoc. Editor (Indianapolis)	Lauren Zeugner	lzeugner@the-papers.com
Notes	Also publishes an edition for Cincinnati, OH.

Senior Life
Address	2120 Second Ave., Terre Haute IN 47807
Telephone	317-641-7759
Telephone (advertising)	812-223-2522
E-mail (news)	andyb@thjournal.com
Web Site	www.wvseniorlife.com
Publication Date	Monthly
Circulation	8,000 (free)
Publishing Company	Flottsom Communications
Publisher	Robert Flott	robertf@thjournal.com
Editor	Timothy Brown	andyb@thjournal.com
Advertising Manager	John Gifford	johng@thjournal.com

SPORTS

Blue & Gold Illustrated

Address	54377 30th St., South Bend IN 46635
Telephone	574-968-1104
Telephone (toll-free)	877-630-8768
Web Site	www.bluegoldonline.com
Publication Date	20 times per year
Circulation	30,000
Publishing Company	Coman Publishing Co.
Editor	Lou Somogyi — lsomogyi@blueandgold.com
Advertising Manager	Michelle Delee-Hamilton — michelle@blueandgold.com
Notes	Covers Notre Dame athletics.

Go Team Magazine

Mailing Address	P.O. Box 14131, Evansville IN 47728
Street Address	4 Chestnut St., Evansville IN 47713
Telephone	812-429-3907
Telephone (toll-free)	866-963-9748
Fax	812-429-3908
E-mail (news)	editor@news-4u.com
Circulation	10,000 (free/newsstand)
Publishing Company	Publications of News4u, Inc.
Publisher	Bashar Hamami — bashar@news-4u.com
General Manager	Sharon Tindle — sharon@news-4u.com
Managing Editor	Dylan Gibbs — editor@news-4u.com
Advertising Manager	Lori Martin — lori@news-4u.com
Sports Editor	Casey McCoy — events@goteammagazine.com
Production Supervisor	Amanda Smith — art@news-4u.com
Notes	Also publishes **Tri-State Bride** and **What's Cookin'! Magazine** (annual publications).

Indianapolis Tennis Magazine

Mailing Address	P.O. Box 21, Bargersville IN 46106
Telephone	317-918-0726
Fax	317-422-4559
Web Site	www.indytennismag.com
Publication Date	Quarterly
Circulation	4,000 (free)
Publishing Company	Cooper Media Group
Publisher	Scott Cooper

Inside Indiana

Mailing Address	P.O. Box 1231, Bloomington IN 47402
Telephone (toll-free)	800-282-4648
Telephone (advertising)	859-619-4575
Fax	812-334-9722
Web Site	www.insideiu.com
Publication Date	Weekly (October-November and January-March)
Circulation	8,000 (paid)
Publishing Company	Landmark Communications
Publisher	Ed Magoni — ed@insideiu.com
Editor	Ken Bikoff — kbikoff@insideiu.com
Advertising Manager	Tim Francis — tfrancis@lcni.com

Irish Sports Report

Street Address	225 W. Colfax Ave., South Bend IN 46626
Telephone	574-235-6428
Fax	574-235-6091
E-mail	isr@sbtinfo.com
Web Site	www.irishsportsreport.com
Publication Date	Weekly (during football season)
Circulation	10,000 (paid)
Publishing Company	Schurz Communications
Publisher	David Ray — dray@sbtinfo.com
General Manager	Steven Funk — sfunk@sbtinfo.com
Editor	Bill Bilinski — bbilinski@sbtinfo.com
Advertising Manager	Carol Smith — csmith@sbtinfo.com
Notes	Covers U. of Notre Dame athletics with emphasis on football and recruiting.

Sports Hotline

Mailing Address	P.O. Box 1183, Marion IN 46952
Telephone	765-664-8732
Fax	765-664-3378
E-mail	kenbhill@bpsinet.com
Publication Date	Weekly (Monday)
Circulation	1,800 (paid)
Publisher	Ken Hill — kenbhill@bpsinet.com
Notes	Covers athletics in Grant County.

WOMEN

HELEN Magazine

Address	658 Main St., STE 215, Lafayette IN 47901
Telephone/Fax	765-429-5297
E-mail	info@helenmagazine.com
Web Site	www.helenmagazine.com
Publication Date	Bi-monthly
Circulation	4,000 (paid)
Publishing Company	Full Moon Communications & Design
Editor	Sharon Martin — sharon@helenmagazine.com
Advertising Manager	Ron Schuessler — hmagazineron@aol.com

Indianapolis Woman Magazine

Address	6610 N. Shadeland Ave., STE 100, Indianapolis IN 46220
Telephone	317-585-5858
Telephone (toll-free)	877-469-6626
Fax	317-585-5855
E-mail (news)	sfinnell@weisscomm.com
Web Site	www.indianapoliswoman.com
Publication Date	Monthly
Circulation	62,100 (controlled)
Publishing Company	Weiss Communications
Publisher	Mary Weiss
Editor	Shari Finnell — sfinnell@weisscomm.com
Director of Sales	Katy Knerr — kknerr@weisscomm.com

NATIONAL RADIO STATIONS SERVING INDIANA

Air 1

Address	2351 Sunset Blvd., STE 170-218, Rocklin CA 95765
Telephone (toll-free)	888-937-2471
Web Site	www.air1.com
Format	Christian Rock
Network Affiliations	Air 1
On-air Hours	24/7
Broadcast Company	Educational Media Foundation
Notes	Broadcasts on 88.5 FM & 103.1 FM (Bloomington) and 88.9 FM (New Albany).

American Family Radio

Mailing Address	P.O. Drawer 2440, Tupelo MS 38803
Street Address	107 Parkgate Dr., Tupelo MS 38801
Telephone	662-844-5036
Fax	662-842-7798
Web Site	www.afr.net
Wattage	500
Format	Religious
On-air Hours	24/7
Broadcast Company	American Family Assn.
General Manager	Marvin Sanders msanders@afr.net
Accepts PSAs?	yes
Notes	Broadcasts on:
	WATI (89.9 FM) serving Vincennes;
	WQSG (90.7 FM) serving Lafayette

K-Love

Address	2351 Sunset Blvd., STE 170-218, Rocklin CA 95765
Telephone (toll-free)	800-434-8400
Web Site	www.klove.com
Format	Contemporary Christian
Network Affiliations	K-Love
On-air Hours	24/7
Broadcast Company	Educational Media Foundation
Notes	Broadcasts on:
	WIKL (90.5 FM) serving Indianapolis;
	WIKV (89.3 FM) serving Plymouth;
	WJLR (91.5 FM) serving Seymour;
	WKHL (106.7 FM) serving Lafayette;
	WKLU (101.9 FM) serving Indianapolis;
	WKMV (88.3 FM) serving Muncie;
	WKVN (95.3 FM) serving Evansville;
	WQKV (88.5 FM) serving Rochester.
	Repeats on: 97.3 FM (Bedford), 102.1 FM (Columbus), 95.1 FM (Floyds Knobs), 97.5 FM (Franklin), 102.9 FM (Muncie), and 101.7 FM (Richmond).

NEWS SERVICES

Associated Press

Address	251 N. Illinois St., STE 1600, Indianapolis IN 46204
Telephone	317-639-5501
Fax	317-638-4611
E-mail	indy@ap.org
Web Site	www.ap.org
Bureau Chief	Keith Robinson

State House Bureau

Telephone	317-631-8629
Correspondent	Mike Smith

South Bend Bureau

Address	223 W. Colfax Ave., South Bend IN 46626
Telephone	574-288-1649
Fax	574-288-3197
Correspondent	Tom Coyne

Hoosier Ag Today

Address	P.O. Box 34236, Indianapolis IN 46234	
Telephone	317-247-9360	
Fax	317-247-9380	
President	Gary Truitt	gtruitt@hoosieragtoday.com
V. P. Operations	Andy Eubank	aeubank@hoosieragtoday.com
Chief Financial Officer	Kathleen Truitt	ktruitt@hoosieragtoday.com
Traffic Manager	Beth Carper	traffic@hoosieragtoday.com
Notes	Agriculture network serving Indiana radio stations.	

Inside Indiana Business with Gerry Dick

Address	1630 N. Meridian St., STE 400, Indianapolis IN 46202
Telephone	317-275-2010
E-mail	newsletter@growindiana.net
Web Site	www.insideindianabusiness.com
Notes	News service covering Indiana business.

Network Indiana

Address	40 Monument Cir., STE 400, Indianapolis IN 46204	
Telephone	317-637-4638	
Fax	317-684-2008	
Web Site	www.networkindiana.com	
General Manager	Charlie Morgan	charliemorgan@indy.emmis.com
Operations Manager	John Emerson	jemerson@emmis.com
General Sales Manager	Eric Wunnenberg	ewunnenberg@wibc.emmis.com
Public Affairs Director	Matt Hibbeln	mhibbeln@wibc.emmis.com
Notes	Network serving Indiana radio stations.	

MICHIGAN

ILLINOIS

OHIO

KENTUCKY

INDIANA

0   40 mi

COUNTY/CITY INDEX

Adams: Berne, Decatur

Allen: Fort Wayne, Grabill, Huntertown, Monroeville, New Haven

Bartholomew: Columbus, Hope

Benton: Fowler, Oxford

Blackford: Hartford City, Montpelier

Boone: Lebanon, Zionsville

Brown: Helmsburg, Nashville

Carroll: Delphi, Flora

Cass: Logansport, Royal Center

Clark: Charlestown, Clarksville, Jeffersonville, New Washington

Clay: Brazil, Clay City

Clinton: Frankfort

Crawford: English, Marengo

Daviess: Odon, Washington

Dearborn: Aurora, Bright, Lawrenceburg

Decatur: Greensburg

DeKalb: Auburn, Butler, Garrett

Delaware: Muncie

Dubois: Ferdinand, Huntingburg, Jasper

Elkhart: Bristol, Elkhart, Goshen, Middlebury, Nappanee, Wakarusa

Fayette: Connersville

Floyd: New Albany

Fountain: Attica, Covington

Franklin: Brookville

Fulton: Kewanna, Rochester

Gibson: Fort Branch, Oakland City, Princeton

Grant: Fairmount, Gas City, Marion, Upland

Greene: Bloomfield, Jasonville, Linton

Hamilton: Arcadia, Carmel, Cicero, Fishers, Noblesville, Westfield

Hancock: Fortville, Greenfield, McCordsville, Mount Comfort, New Palestine, Wilkinson

Harrison: Corydon, Palmyra

Hendricks: Avon, Brownsburg, Danville, Plainfield

Henry: Knightstown, Middletown, New Castle

Howard: Kokomo

Huntington: Huntington, Warren

Jackson: Brownstown, Crothersville, Seymour

Jasper: DeMotte, Remington, Rensselaer

Jay: Dunkirk, Portland

Jefferson: Madison

Jennings: North Vernon

Johnson: Edinburgh, Franklin, Greenwood, Whiteland

Knox: Bicknell, Vincennes

Kosciusko: Milford, Syracuse, Warsaw

LaGrange: Howe, LaGrange

Lake: Cedar Lake, Crown Point, Dyer, Gary, Hammond, Hobart, Lowell, Merrillville, Munster, Whiting, Winfield

LaPorte: LaPorte, Michigan City

Lawrence: Bedford, Mitchell

Madison: Alexandria, Anderson, Elwood, Lapel, Pendleton

Marion: Beech Grove, Indianapolis, Speedway

Marshall: Bourbon, Bremen, Culver, Plymouth

Martin: Loogootee, Shoals

Miami: Peru

Monroe: Bloomington, Ellettsville

Montgomery: Crawfordsville

Morgan: Martinsville, Mooresville

Newton: Brook, Kentland, Morocco

Noble: Albion, Avilla, Kendallville, Ligonier

Ohio: Rising Sun

Orange: French Lick, Orleans, Paoli

Owen: Spencer

Parke: Rockville

Perry: Tell City

Pike: Petersburg

Porter: Chesterton, Hebron, Portage, Valparaiso

Posey: Mount Vernon, New Harmony

Pulaski: Francesville, Winamac

Putnam: Cloverdale, Greencastle

Randolph: Union City, Winchester

Ripley: Batesville, Versailles

Rush: Rushville

Saint Joseph: Granger, Mishawaka, Notre Dame, South Bend, Walkerton

Scott: Austin, Scottsburg

Shelby: Flat Rock, Morristown, Shelbyville

Spencer: Dale, Rockport, Santa Claus

Starke: Knox

Steuben: Angola, Hamilton, Orland

Sullivan: Sullivan

Switzerland: Vevay

Tippecanoe: Lafayette, West Lafayette

Tipton: Tipton

Union: Liberty

Vanderburgh: Evansville

Vermillion: Cayuga, Clinton

Vigo: Terre Haute

Wabash: North Manchester, Wabash

Warren: Williamsport

Warrick: Boonville, Newburgh

Washington: Hardinsburg, Pekin, Salem

Wayne: Cambridge City, Centerville, Hagerstown, Richmond

Wells: Bluffton, Ossian

White: Monon, Monticello, Wolcott

Whitley: Churubusco, Columbia City, South Whitley

INDIANA RADIO STATIONS

WABX - 51	WBZQ - 67	WFRR - 45	WIRE - 127	WLAB - 70
WADM - 43	WCBK - 137	WFWI - 66	WISU - 190	WLBC - 150
WAJI - 64	WCDQ - 40	WFWR - 9	WITT - 104	WLDC - 80
WAKE - 196	WCFY - 52	WFYI - 102	WITZ - 112	WLDE - 64
WAMW - 207	WCJC - 135	WGAB - 54	WIUX - 19	WLEG - 46
WAOR - 143	WCJL - 197	WGBF - 53	WIVR - 114	WLFQ - 46
WAOV - 200	WCKZ - 65	WGBJ - 68	WIWC - 7	WLFW - 51
WARU - 163	WCLS - 17	WGCL - 19	WIWU - 136	WLHK - 101
WASK - 121	WCMR - 45	WGCS - 79	WJAA - 177	WLHM - 130
WATI - 235	WCOE - 125	WGL - 68	WJCF - 83	WLJE - 196
WAUZ - 33	WCRD - 149	WGLL - 11	WJCO - 197	WLKI - 8
WAWC - 205	WCRT - 189	WGNR - 7	WJCP - 157	WLME - 186
WAWK - 113	WCSI - 34	WGRE - 81	WJCY - 197	WLOI - 125
WAXI - 188	WCVL - 40	WGVE - 78	WJEF - 122	WLPR - 138
WAXL - 90	WCYT - 67	WHBU - 150	WJEL - 105	WLQI - 168
WAZY - 122	WDKS - 53	WHCC - 17	WJFX - 66	WLTH - 139
WBAA - 208	WDND - 180	WHFB - 144	WJHS - 32	WLYV - 70
WBAT - 135	WDSO - 29	WHHH - 103	WJJD - 119	WMBL - 7
WBCL - 65	WEAX - 8	WHJE - 26	WJJK - 100	WMDH - 154
WBDC - 90	WECI - 169	WHLP - 197	WJLR - 235	WMEE - 69
WBDG - 98	WEDJ - 98	WHME - 181	WJLT - 53	WMGI - 190
WBEW - 28	WEDM - 99	WHOJ - 189	WJOB - 86	WMHD - 191
WBEZ - 28	WEEM - 162	WHON - 171	WJOE - 66	WMPI - 176
WBGW - 52	WEFM - 140	WHPD - 181	WJOT - 203	WMQX - 150
WBHW - 52	WEJK - 51	WHPL - 7	WJPR - 131	WMRI - 135
WBIW - 14	WENS - 83	WHPZ - 181	WJYW - 194	WMRS - 147
WBKE - 157	WEOA - 54	WHUM - 35	WKAM - 79	WMXQ - 150
WBNI - 65	WERK - 150	WHWE - 89	WKBV - 170	WMYJ - 20, 137
WBNL - 22	WETL - 180	WHZR - 130	WKDQ - 53	WMYK - 118
WBOI - 65	WFBQ - 99	WIBC - 104	WKHL - 235	WMYQ - 67
WBOW - 188	WFCI - 75	WIBN - 159	WKHY - 121	WNAS - 153
WBPE - 122	WFCV - 67	WIBQ - 190	WKID - 198	WNDA - 30
WBRI - 98	WFDM - 76	WICR - 104	WKJG - 69	WNDE - 99
WBRO - 133	WFGA - 10	WIFE - 36, 175	WKKG - 34	WNDI - 185
WBSB - 149	WFHB - 18	WIKI - 132	WKLO - 160	WNDV - 180
WBSH - 149	WFIU - 18	WIKL - 235	WKLU - 235	WNDY - 40
WBSJ - 149	WFLQ - 77	WIKV - 235	WKMV - 235	WNDZ - 165
WBST - 149	WFMG - 170	WIKY - 51	WKOA - 121	WNHT - 68
WBSW - 149	WFML - 201	WILO - 74	WKPW - 115	WNIN - 54
WBTO - 200, 207	WFMS - 100	WIMC - 40	WKRY - 33	WNOU - 103
WBTU - 66	WFNI - 101	WIMS - 141	WKUZ - 203	WNRL - 128
WBWB - 17	WFOF - 7	WINN - 34	WKVI - 116	WNSN - 144
WBYR - 66	WFRI - 45	WIOE - 206	WKVN - 235	WNTR - 105
WBYT - 143	WFRN - 45	WIOU - 118	WKZS - 38	WNTS - 106

WNUY - 70	WRBR - 143	WSJD - 167	WUEV - 55	WXRD - 196
WOCC - 38	WRCY - 148, 200	WSKL - 38	WUME - 160	WXXB - 121
WOJC - 197	WRDZ - 27	WSLM - 175	WUZR - 200	WXXC - 135
WORX - 133	WREB - 81, 200	WSMM - 180	WVHI - 55	WYCA - 87
WOWO - 69	WRFM - 83	WSND - 158	WVLP - 197	WYFX - 148, 200
WPFR - 191	WRFT - 106	WSPM - 106	WVNI - 20	WYGB - 35
WPGW - 166	WRGF - 83	WSRB - 87	WVPE - 47	WYGS - 33
WPHZ - 14	WRIN - 168	WSTO - 51	WVRG - 40	WYIR - 56
WPSR - 55	WROI - 173	WSVX - 178	WVSH - 91	WYPW - 143
WPUM - 168	WRSW - 205	WSWI - 55	WVUB - 202	WYTJ - 128
WPWX - 87	WRWM - 100	WSYW - 98	WVUR - 197	WYXB - 101
WQHK - 69	WRZQ - 35	WTCA - 164	WWBL - 200, 207	WZBD - 15
WQHU - 91	WRZR - 131	WTCJ - 186	WWCA - 139	WZDM - 200
WQKO - 197	WRZX - 99	WTGO - 123	WWCC - 123	WZOC - 144
WQKV - 235	WSAL - 130	WTHD - 124	WWKI - 119	WZOW - 180
WQKZ - 112	WSBL - 181	WTHI - 192	WWSY - 190	WZPL - 105
WQLK - 171	WSBT - 144	WTLC - 103	WWVR - 192	WZVN - 196
WQME - 7	WSCH - 126	WTMK - 197	WWWY - 34	WZWZ - 118
WQRK - 14	WSDM - 188	WTRC - 46	WXFN - 150	WZZB - 177
WQSG - 235	WSDX - 188	WTRE - 84	WXGO - 133	WZZY - 170
WQSW - 71	WSEZ - 160	WTTS - 19, 106	WXKE - 68	XRB - 25
WQTY - 200	WSHP - 122	WTUR - 194	WXKU - 177	
WRAY - 166	WSHW - 74	WUBS - 182	WXLW - 76	
WRBI - 13	WSHY - 122	WUBU - 182	WXNT - 105	

INDIANA TELEVISION STATIONS

Community Access Television Services - 20
HomeTown Sports Indiana - 156
HomeTown Television - 156
TV3 - 37
W07CL - 11
WANE - 71
WAZE - 56
WBND - 145
WCLJ - 85
WCTV - 171
WCWW - 145
WDNI - 107
WEHT - 56
WEVV - 57
WFFT - 71
WFIE - 57
WFWA - 72
WFWC - 11
WFXW - 192
WFYI - 107
WHAN - 175
WHMB - 156
WHME - 182
WHNW - 182
WINM - 72
WIPB - 151
WIPX - 107
WISE - 73
WISH - 108
WIWU - 136

WJTS - 112
WJYL - 30
WKOG - 108
WKOI - 172
WLFI - 208
WMUN - 151
WMYS - 145
WNDA - 30
WNDU - 183
WNDY - 108
WNIN - 57
WNIT - 47
WPTA - 73
WREP - 138
WRTV - 109
WSBT - 145
WSJV - 183
WSOT - 136
WTHI - 193
WTHR - 109
WTIU - 21
WTSN - 58
WTTK - 110
WTTV - 110
WTVW - 58
WTWO - 192
WVUT - 202
WXIN - 110
WYCS - 30
WYIN - 139

INDIANA NEWSPAPERS & PUBLICATIONS

MEDIA DIRECTORY ORDER FORM

Name _____

Title _____

Company _____

Street Address _____

City, State and Zip _____

Mailing Address (if different than above) _____

City, State and Zip _____

Telephone _____ Fax _____

E-mail _____

PLEASE SEND ME THE 2010 INDIANA MEDIA DIRECTORY (available now):

_____ books @ $85 each $_____

PLEASE SEND ME THE 2011 INDIANA MEDIA DIRECTORY (available early 2011):

_____ books @ $85 each $_____

SUB-TOTAL	$_____
7% INDIANA SALES TAX	+ $_____
TOTAL ORDER	**TOTAL** $_____

SHIPPING: Brackemyre Publishing is celebrating its 23rd anniversary by offering **FREE SHIPPING** on all products! All orders will be shipped within 3 business days of receipt of order.

METHOD OF PAYMENT:

❐ Check (payable to Brackemyre Publishing) enclosed.

❐ Bill me.   Signature required   X _____

❐ Please charge to MasterCard or VISA:

     Cardholder's Name _____

     Card Number _____

     Expiration Date _____

     Signature Required _____

BRACKEMYRE PUBLISHING

Mail:	10133 Preston Court, Fishers IN 46037	**Phone:**	317-913-1655
E-mail:	mediadirectory@comcast.net	**Fax:**	317-598-2609

www.MyMediaDirectory.Com